Planning as Persuasive Storytelling

New Practices of Inquiry

A series edited by John S. Nelson and D. N. McCloskey

Planning as Persuasive Storytelling: The Rhetorical Construction of Chicago's Electric Future

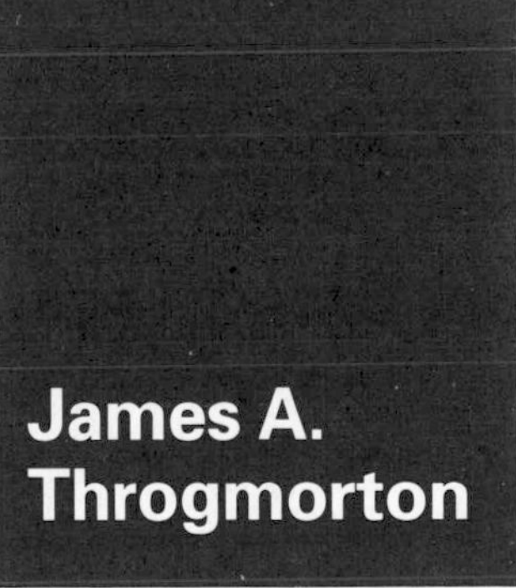

THE UNIVERSITY OF CHICAGO PRESS CHICAGO & LONDON

James A. Throgmorton is associate professor in the Graduate Program in Urban and Regional Planning at the University of Iowa.

The University of Chicago Press, Chicago 60637
The University of Chicago Press, Ltd., London

Printed in the United States of America
05 04 03 02 01 00 99 98 97 96 1 2 3 4 5
ISBN: 0-226-79963-8 (cloth)
0-226-79964-6 (paper)

Library of Congress Cataloging-in-Publication Data
Throgmorton, James A.
Planning as persuasive storytelling : the rhetorical construction of Chicago's electric future / James A. Throgmorton.
p. cm.—(New practices of inquiry)
Includes bibliographical references and index.
1. Commonwealth Edison Company. 2. Electric utilities—Illinois—Chicago Metropolitan Area. 3. Electric power—Illinois—Chicago Metropolitan Area—Planning. I. Series.
HD9685.U7C698 1996
333.79′32′097311—dc20 95-42963
CIP

♾The paper used in this publication meets the minimum requirements of the American National Standard for Information Sciences—Permanence of Paper for Printed Library Materials, ANSI Z39.48-1984.

Contents

Illustrations

Tables

Preface

We citizens repeatedly hear that scientific planning can and should replace politics, particularly in highly technical arenas such as the electric power industry. We hear, for example, that scientists should determine how and where to store highly radioactive nuclear wastes most safely. Conditioned to want scientific criteria to guide such choices, we become deeply frustrated and perhaps resentful when we observe political interests dominating the scientific planners' inquiries. The president announces—only weeks before a major national election—that the new super conducting supercollider will be located in Texas? There the politicians go again, we are likely to respond. Frustrated and resentful, we tend to throw up our hands and condemn politicians for being corrupt and scientists for being hired guns at worst and naive fools at best. A plague on both their houses.

This book can be thought of as an extended reaction against (and commentary on) this modernist conception of planning and on the effort to "modernize" cities and the functional systems (including electric power) that serve them. As such, it addresses four major interconnected topics:

- the relationship between science and politics
- the relationship between planning and modernization
- structural transformation away from regulated natural monopoly in the electric power industry
- the importance of persuasive and constitutive storytelling in contemporary planning.

Rather than insisting that "scientific planning" should replace "politics," I try to articulate, justify, and exemplify an alternative understanding of planning that is better suited to the fragmented and conflictual nature of life and politics in contemporary America, and to help planners and others learn how to argue coherently and persuasively about contestable views of what is good, right, and feasible. According to the alternate view that I put forth, planning can best be thought of as a form of persuasive and constitutive storytelling that occurs within a web of relationships and partial truths. In this view, planners can be regarded as authors who write future-oriented texts (plans, analyses, articles) that reflect awareness of differing or opposing views and that can be read (constructed and interpreted) in diverse and often conflicting ways. But planners do not simply write texts; they are also characters whose forecasts, surveys, models, and other tools act as tropes (persuasive figures of speech and argument) in the planning stories that they and others tell.

The book seeks to justify this alternative understanding by weaving an interconnected web of stories. The web begins with a brief history of planning in America, then spirals inward through three more levels: about electric power planning as practiced by the Commonwealth Edison Company (Edison) and regulated by the Illinois Commerce Commission (ICC), about the City of Chicago's efforts to escape (or at least alter the terms of) that electric power planning and regulatory system, and about one survey researcher's efforts to persuade a citizens task force that he had truthfully measured the attitudes and likely responses of Chicago businesspeople to the city's efforts.

That spiraling inward can be seen quite clearly in the book's overall structure. Chapter 1 shows planning beginning as a modernist enterprise in which planners would have the knowledge and power to guide the development of cities. As that chapter unfolds, however, it shows planning being transformed to a point where planning theorists now offer conflicting and often contradictory accounts about what planning is and what its object of attention should be. Chapter 2 suggests that the presence of those conflicting and contradictory accounts have inspired a rhetorical or argumentative turn in planning. Accordingly it draws upon an eclectic body of literature, some from within planning but most not, to construct a rhetorical understanding of planning and to argue that planning can be thought of as a form of persuasive and constitutive storytelling about the future. Having presented in chapters 1 and 2 a rationale for attending to persuasive storytelling, the book then illustrates the importance of such discourse by narrating (in chapters 3 through 10) an intricate tale about electric power planning in the Chicago area from 1973 through 1993. This detailed story takes place at three separate levels, weaving if you will ever deeper into the rhetorical dimensions of planning. Chapter 3 quickly states the modernist approach to electric-power planning and regulation (that is, the "normal discourse" of public utility economics), and chapter 4 tells how

Commonwealth Edison relied on modernist planning to adopt—and from 1973 through 1986 to implement—the nation's most ambitious nuclear power expansion plan. Chapters 5, 6, and 10 recount Edison's unsuccessful efforts between 1986 and 1993 to persuade relevant parties (consumer groups, the ICC, the Illinois Supreme Court, and others) to approve rate increases needed (in Edison's judgment) to pay for the last three nuclear units in its expansion plan. The book shifts to Chicago in chapters 7 through 9. There it recounts the City of Chicago's hesitant efforts from 1985 through 1991 to explore alternatives to remaining on Edison's electric power system, or at least to negotiate more favorable contractual terms with the utility. In some ways, chapter 8 forms the heart of the book. It tells of a survey researcher's effort in 1989 to persuade a mayorally appointed citizens task force that he had truthfully measured the likely response of Chicago businesspeople to a city "takeover" of part of Commonwealth Edison's system. The book concludes in chapter 11 by returning to the themes introduced at the start and by commenting on the relevance of the electric power planning story for planning in contemporary America.

The book also seeks to exemplify its claims about planning by *being* a persuasive and constitutive story about the future. The presumed persuasiveness comes in part from the book's reliance on a variety of stylistic devices, including a prelude and interludes between chapters, authorial inserts, photographs, maps, a painting, a venn diagram, advertisements, twelve line graphs, and extended technical commentaries. The prelude, ten interludes between chapters, and the postlude seek to draw readers into the web, to help them feel the settings in which important events take place, and to introduce them to the real flesh and blood author and what he feels and thinks when in those settings. At another level, the interludes demonstrate at least two ways in which the author is part of the web: as the peripheral first person narrator of the electric power tales, and as a theorist, teacher, and practitioner of planning.

I use over forty authorial inserts primarily to comment on statements that someone else makes in the book, seeking to convey the point that dialogue and conversation are vitally important parts of persuasiveness. In addition, the inserts often introduce (without elaboration) issues that emerge later in the book, or harken back to themes that were introduced earlier and thereby help tie the otherwise rather complicated book together.

This is what an authorial insert looks like. Many readers might find the inserts intrusive at first, but—alerted to expect them—are likely to find them of increasing value as the book proceeds.

I present more than twenty photographs, usually to bring to life the characters that appear in the story. The photographs themselves have been chosen so as to accentuate an important aspect of each character. In chapter 6, for example, Chairman Barnich appears to be a handsome, confident, and successful young man. Much later, in chapter 9, Mayor Daley looks simultaneously bored and exasperated as he and Alderman Gutierrez draw an important negotiation and compromise to a close.

I also present two maps, both of which express visually the modernist view that the city is like a machine and that planners are like the engineers and technicians who make it and its functional systems run efficiently; a painting which views the modernist planners' gleaming towers from the teeming slums; a venn diagram to express my claim that planning draws from the three core impulses of politics, science, and advocacy; several of Edison's television, radio, and newspaper advertisements; and twelve line charts that document projected and actual trends in Edison's sales, peak load, earnings, and revenue per unit of electricity sold. Lastly, I present five extended technical commentaries in chapter 4. They are too long and too important for footnotes, yet too peripheral to the main story line to be brought directly into the text. Set apart in boxes, they help readers see connections between what was happening in the Chicago area and what was happening at Three Mile Island, Chernobyl, and other parts of the web.

The topics, structure, and expositional devices are likely to interest and challenge at least five distinct audiences. Scholars of planning, policy analysis, and policy making are quite likely to be interested and challenged by the book's contribution to the theories and practices of planning and policy making. Professional city planners might at first respond with some ambivalence. Though they will probably find the history of urban and regional planning in chapter 1 to be comfortably familiar, they might feel that the five chapters about electric power planning and regulation lead well away from their work. On the other hand, these practitioners might well be fascinated by chapters 7 through 9, and they might find my effort to explicate the labyrinthine nuances of electric power planning rhetoric to be interesting, enlightening, and quite relevant to the intricacies of their own rhetoric. The book is quite likely to attract the interest of engineers, lawyers, economists, and others who are directly involved in electric power planning, policy making, and regulation. The story of Edison's difficulty in completing its nuclear expansion plan on time and within budget, in obtaining "timely rate relief," and in trying to shape the outcome of Chicago's exploration of options, is a story that those practitioners are likely to find important, compelling, and discomfortingly familiar. I expect them to read the book and to react strongly (and in diverse ways) to it. I also expect that the book will be of considerable interest to people who live in northern Illinois, particularly Chicago. They have very strong feelings about Edison, about the City of Chicago, and about regulatory politics in their state, so they too will probably read this book with passionate interest. Lastly, I have strived to attract and retain the interest of the large numbers of American citizens who are concerned about the politics and practices of their governments but who would be alienated by the normal discourses of public utility economics, planning, and law. They routinely hear that planning has failed, just as "centralized bureaucracy" and the "welfare state" have failed, but they also cry out for good planning in their own local communities. I have sought to write the book in such a way as to create a temporary community with and among all five of those audiences. So the book seeks to be constitutive as well as persuasive.

A brief word about the author would be appropriate, given the approach to planning that I set forth. I feel as though I am but a minor character in a long and extremely important story about planning in America. Minor though I feel, I still have something important to say, something to contribute to the continuing conversation about what it means to act intelligently and wisely with regard to the future. In this connection, it is vitally important to stress that I wrote this book not as a well-informed and deeply committed insider, and not as an ideal objective observer, but as a peripheral first person narrator. Other people in the Chicago area know particular details of the story far better than I do, and other scholars can put a much finer point on particular details. My contribution is to weave it all into a well-integrated story and to construct a story as seen from the point of view of a planning scholar.

FIG. 1 **The Author**

As a real flesh and blood author, I should also report that I have been affected both personally and professionally by the modernist planners' efforts to reshape the world. The planting of the seed for this book can be traced back to 1985 when I was working as an energy policy analyst for Argonne National Laboratory near Chicago. A government-owned, contractor-operated laboratory, Argonne primarily conducted energy-related research for the U.S. Department of Energy (DOE). At the time I was managing a research project for one of DOE's offices. Given the enabling legislation which created that office and my own interests in national electric utility policy, I was curious about the potential effects of DOE's energy policies, particularly those concerning nuclear power, on the health and economic livelihood of minority populations. Perhaps somewhat naive, I was surprised to discover important political constraints on my research team's ability to explore those effects. Three particular instances of those constraints impressed me most. We proposed to study the possible economic effects of nuclear power plant abandonments on minority populations, but—fearing adverse reaction from some electric utilities—DOE officials instructed us not to pursue the study. Similarly we were told not to investigate the potential costs and benefits of "least-cost utility planning" for minorities (see Throgmorton and Bernard 1986 and Throgmorton 1989). Lastly we proposed to contribute technical analyses relating to the City of Chicago's effort to explore alternatives to remaining on Commonwealth Edison's electric power system, but Argonne management declined to let us work with the city. What impressed me most was the conflict between the DOE's and Argonne's obvious (and perhaps quite necessary) political behavior and their public claims to be sponsoring and conducting unbiased, objective, scientific research. Thus, this book grows partly out of my personal experiences of science interacting with politics in a highly technical arena.

In some ways, however, the origins of this study lie farther back in time. My doctoral dissertation at the University of California at Los Angeles School of Architecture and Urban Planning, which I completed in 1983, investigated the implementation of Section 210 of the Public Utilities Regulatory Policies Act of 1978. That research persuaded me that the electric utility industry was just beginning to move away from what I refer to in the present book as the institution of regulated natural monopoly. The crucial questions were what direction would that change take, and why?

Even deeper roots lie in the fact that my mother and father met and married while they were working for the Tennessee Valley Authority (TVA) in 1939 and that I was born while my father was working as a supervisor of construction at Kentucky Dam near Gilbertsville, Kentucky. Thus I am almost literally a child of the TVA and the electric power reforms of the Roosevelt Administration. As a child I was affected by the romance of dam building, and I have always admired and been deeply impressed by my father's ability to build. As an adult living in different times and confronting different circumstances than my father, however, I have felt a strong need to help build something other than dams and other massive engineering marvels that impose themselves on the natural and social world. I have sought to discover or help create a mode of planning that is based on "being-in-the-world" rather than being apart from it. The idea that planning can be conceived of as persuasive and constitutive storytelling within a web of relationships is a product of that search.

Acknowledgments

The approach to planning which I argue for in this book (persuasive storytelling within a web of relationships) is based in large part on the web of relationships of which I am a part. I do not consider myself an atomized individual, and I do not consider this book to be a product of my isolated, contemplating mind. What I think and have written is very much a product of interaction with other people, some of them scholars, some of them energy professionals in the Chicago area, and some of them personal friends. "I" acknowledge my debts to them.

Let me mention just a few of the relationships that have affected my writing most intensely, knowing that there are others that will unintentionally go unacknowledged. I was a customer of the Commonwealth Edison Company from 1983 through 1986. As a customer, I chose to become a member of the Citizens Utility Board in 1984, and I remained a member until I left Illinois in 1986. Though living in Iowa from that year to the present, I chose to become a member of the Chicago Electric Options Campaign in 1989 and 1990, and I have contributed small amounts of money to the Center for Neighborhood Technology. Wanting to gain a deeper understanding of the electric utility's perspective, and hoping to earn a modest return on my capital, I purchased a small number of common shares in the Commonwealth Edison Company in 1990. Each of these relationships has affected who I am and how I think and write.

I have also been deeply affected by several scholars, professionals, and community activists, and I want to acknowledge their advice, inspiration, and openness to my curious mind and eye. I include in this list several friends at Argonne (including Mike Meshenberg, Ed Tanzman, and Dee Wernette); people active in electric power planning in the Chicago area (including Scott Bernstein, Maureen Dolan, Alderman Edwin Eisendrath, and Charles Williams); scholars thoughout the United States (including Ernest Alexander, William Ascher, Peter

Fisher, John Forester, John Friedmann, Sue Hendler, Charlie Hoch, Jack Kartez, Jimmie Killingsworth, Helen Liggett, Peter Marris, Seymour Mandelbaum, Dowell Myers, Mike Sheehan, Marty Wachs); anonymous reviewers from the National Science Foundation, several scholarly journals, and the University of Chicago Press; participants in the University of Iowa's 1990 Conference on Narrative in the Human Sciences and its Spring 1991 Scholars Workshop on the Rhetoric of Political Argumentation (including Fred Antczak, Ed Arrington, Bob Boynton, and Kathleen Farrell); participants in annual meetings of the Association of Collegiate Schools of Planning from 1987 through 1992 (including the Joint International Planning Congress held in Oxford, England, in 1991); and attendees at the 1991 Conference on the Discourse of Environmental Advocacy. I also wish to acknowledge the support services provided by Jay Semel and Lorna Olsen at the Center for Advanced Study at the University of Iowa and the editorial assistance provided by Doug Mitchell and his colleagues at the University of Chicago Press, and to thank the many diverse members of the University of Iowa's Project on the Rhetoric of Inquiry, including the Project's executive director, Kate Neckerman. Most importantly, I want to extend a word of admiration and appreciation to John Nelson of the University of Iowa. What a tireless and selfless scholar! I cannot imagine what this book would be like, indeed whether it would even exist, were it not for my extended conversation with him over the past eight years.

As the acknowledgments just presented indicate, this book does not appear out of thin air. Indeed several of its chapters represent modified versions of articles or book chapters that have appeared previously. Chapter 2 revises and expands on three separate journal articles of mine: "Passion, Reason, and Power: The Rhetorics of Electric Power Planning in Chicago," *Journal of Architectural and Planning Research* 7 (1990): 331–50; "The Rhetorics of Policy Analysis," *Policy Sciences* 24 (1990): 153–79; and "Planning as Persuasive and Constitutive Storytelling about the Future: Negotiating an Electric Power Rate Settlement in Illinois," *Journal of Planning Education and Research* 12 (1992): 17–31. Chapter 5 derives primarily from the *Journal of Planning Education and Research* article. Chapter 7 revises material that previously appeared in my *Journal of Architectural and Planning Research* article, but also draws from a piece that I coauthored with Peter S. Fisher: "Institutional Change and Electric Power in the City of Chicago," *Journal of Economic Issues* 27 (1993): 117–52. It also derives from a chapter of mine ("Ethics, Passion, Reason, and Power: The Rhetorics of Electric Power Planning in Chicago") that appears in a book recently edited by Sue Hendler (1995): *Planning Ethics: A Reader in Planning Theory, Practice, and Education*, published by CUPR Press. Chapter 8 revises material that previously appeared in "Planning as a Rhetorical Activity: Survey Research as a Trope in Arguments about Electric Power Planning in Chicago," *Journal of the American Planning Association* 59 (1993): 334–46. Variants of this chapter have also appeared in a book edited by Frank Fischer and John Forester (1993): *The Argumentative Turn*, published by Duke University Press; and in a book edited by Seymour Mandelbaum, Luigi Mazza, and Robert Burchell (forthcoming): *Explorations in Planning Theory*, published by CUPR Press. Lastly, chapter 9 significantly revises the material that previously appeared in the *Journal of Economic Issues* article which I coauthored with Peter S. Fisher.

I should also acknowledge that this book is based in part on work supported by the National Science Foundation's Ethics and Values branch under Grant No. DIR–8911870, and that the government has certain rights in this material. It was not, however, undertaken by or on behalf of the Foundation or any other agency of the U.S. government.

More than any other person mentioned thus far, however, I want to thank my fine young sons Patrick and Paul for being themselves and for helping me to experience life again with beginner's eyes.

Prelude

A Strange Place, an Alien Language

Upon entering the room, we see seven smartly dressed men and women seated behind an elongated table raised slightly off the floor. Another six to eight people face them, three or four sitting behind one table and the others behind a second. A podium stands between these last two tables, about ten feet in front of the elevated platform, and it seems to link the three separate groups. Another fifty or so equally well-attired people sit attentively in some ten rows of folding metal chairs arrayed neatly behind the leading pair of tables.

A handsome, mid-thirties man at the center of the head table speaks first: "This is a special open meeting of the Illinois Commerce Commission pursuant to notice of the Illinois Open Meetings Act. The purpose is to hear oral argument on Docket 90-0169, Proposed General Increase in Electric Rates for the Commonwealth Edison Company. All Commissioners are present." He refers to a long agenda that gives the "parties" specific amounts of time to speak. We learn that the parties include "Edison," "Staff," "BPI," and others. Their "counsels," who sit at the tables adjacent to the podium and in the rows of metal chairs behind them, then introduce themselves and their organizations. After discussing briefly whether this "docket" should be consolidated with Docket 87-0427, the handsome young man then suggests that they begin "oral argument."

Edison's counsel argues first. He is bald, graying on the side, and impeccably dressed in a gray suit. We listen with considerable puzzlement to this older man as he refers to the Illinois Public Utilities Act and basic legal principles, basic economic theory, the cost-recovery standard, just and reasonable rates, a reasonable rate of return on common equity, rate design, monopoly supply characteristics, risk, confiscation of utility property, the hearing examiner's proposed order, the intervenors, prudently incurred and reasonable costs, affordable rates, revenue per kilowatt-hour, earnings per share, and the Supreme Court's BPI decision. Our sleepy-eyed confusion increases as we then hear a younger man, bearded and balding, rise to present BPI's response to the Edison counsel's opening remarks. The younger counsel challenges the Edison argument, but he speaks the same language. He insists that Edison should charge consumers electric power rates that are affordable. That, at least, we can understand. But mostly he talks about plants that are used and useful, about "our brief," the AFUDC, write-offs, the rollback, balancing consumer versus investor interests, deferred charges, the test year, the Sixth Interim Order, and Byron 2 and Braidwood 1. We try to listen intently, but mainly we hear that we are in a strange place where people—who seem to understand one another well enough—speak a language that is largely alien to us.

"Who are these people?" you whisper to me. "Why are they arguing in such measured tones in this particular place and time? What is this strange language they speak? What stories are they telling one another?"

"Good questions," I whisper back. "It's more a matter of the story that they are living with one another. It's a complicated and proverbially long story. Perhaps we can talk about it as we drive back to Iowa City. For now, though, let me just remind you what you already know: they are electric power people; and like all specialists, they are speaking the jargon of their trade. They oppose one another on specific issues; still, they speak a common language. They have shared many experiences, and so they share many understandings with one another. Their arguments concern decisions of the Illinois Commerce Commission and the courts. They are all trying to persuade the ICC commissioners to override their opponents' objections. And they do this by appealing to agreed-upon conventions in the language they call public utility planning and regulation.

"I was with you for a moment, but now you're starting to sound like a specialist, talking in jargon. I'm more confused than ever, even though I know we started to discuss this stuff on our way here."

"You're right. Let's talk about it more in the car."

1

The Irony of Modernist Planning

> Trapped indoors by the onset of winter, and having time to meditate, René Descartes began to compose his *Discourse on Method* (1637). In it he observed that "often there is less perfection in works composed of several separate pieces and made by different masters, than in those at which only one person has worked. So it is that these old cities . . . are usually so badly proportioned in comparison with those orderly towns which an engineer designs at will on some plain that . . . one has the impression that they are more the product of chance than that of a human will operating according to reason" (Megill, 1989).

Think then act. Climb the "central plateau" from which an ideal, objective observer can discern what is best for society. Standing there, be confident that disciplined forethought will make society's collective actions more rational, its judgments more principled, and its future more subject to human control. Forget or "creatively destroy" the past, treating existing laws, institutions, environmental conditions, and patterns of power and influence as irrelevancies at best and obstructions at worst. Wipe the slate clean and "make it new," all in the name of Progress, Order, and Reason.

To the contemporary ear, these words might sound anachronistic or naive or threatening. But no matter how much planners might want to distance themselves from them, the words still provide the foundational principles of planning practice (at least as it is commonly understood). And for good reason: between 1907 and 1917, the core idea that those words express—that *scientific planning should guide societal change*—became embedded in state public utility commissions, city planning commissions, the Federal Reserve Board, and other related public institutions.[1] And then in the 1950s and 1960s, that core idea provided the bedrock for graduate-level planning education.

Though still molded by the ideal of standing on the central plateau, most practitioners and theorists have learned that planners do not occupy such an enlightening or noble or powerful place. Chastened by their evident inability to manage modern economies, guide urban development, or solve long-standing social problems, they have been turning away from (or should I say, coming down from) the plateau since the mid–1960s. Duly humble, they have been trying to articulate new definitions of planning and new roles for planners. No longer seeing themselves as scientific (rational) planners, they have alternately characterized themselves as incrementalists, advocate planners, radical planners, policy analysts, strategic planners, mediators, and so on.

Though rejecting the totalizing discourse of rational planning, many late modern planners still firmly believe that scientific planning can make public policies and cities more rational. Shaped by this principle, they confront a persistent puzzle; namely, the oft-lamented refusal of decision makers to adopt and faithfully pursue the planners' recommendations. Time and again, planners find their advice ignored by specific decision makers, at specific times, in specific places, on specific issues. Time and again, the "ignorant masses" (usually disparagingly referred to as interest groups) carry the day. Time and again, politics (a code word for venality, corruption, and the intervention of know-nothings) defeats the scientific planners' efforts to make cities and related public policies more rational.

To increase public rationality, many contemporary planners tell themselves that they must think scientifically but act politically. Immersed in a highly politicized world in which their opponents do not hesitate to lie, distort, deceive, and act strategically and manipulatively, and concluding that planners must do the same in order to survive and succeed, they increasingly seek to redefine planning as a practice that ultimately rests on bargaining, negotiating, and mediating among competing interests.

Most of these late-modern planners see no significant conflict between the underpinnings and the practice of their profession. Other planners, however, see deep conflict, and have recommended an argumentative or rhetorical turn in planning thought (e.g., Goldstein 1984; Mandelbaum 1988; Forester 1989; Beauregard 1989; Mandelbaum

1990; Hoch 1990; Beauregard 1991; Hillier 1991; Tett and Wolfe 1991; Healey 1992; Healey 1993; Innes 1995). I seek to contribute to that rhetorical turn by arguing that *good planning is persuasive and constitutive storytelling about the future.* In this view, planners should surrender any further pretense to the neutrality, objectivity, and universal Truth and Goodness that modernists hoped to find on the central plateau. Surrendering the pretense to objectivity does not mean, however, that planners should flee to the extreme of defining planning as just another form of politics gone amok. They should, instead, embrace the idea that planning is scientific *and* political, technical *and* persuasive, and that the "tools" planners use act as tropes (persuasive figures of speech and argument) in the planning stories that they tell.

In order to justify making this rhetorical turn, I must first construct a history of planning in America that makes the turn seem plausible and potentially meritorious. I must, in other words, fit this book and the rhetorical turn that it advances into the larger context of planning history. Notice that I must *construct a* history rather than *write the* history. If good planning is persuasive and constitutive storytelling about the future, then it follows that good planning history is persuasive and constitutive storytelling about the past. Thus the history of planning in America does not simply exist, out there in past time, waiting for someone like me to uncover and reveal it; rather, that history can be known only insofar as it is constructed and narrated by fallible authors who write as members of a larger community of readers. Thus the act of constructing a history of planning in America is simultaneously an act of constructing (at least temporarily) a community of readers.

Let me provide three examples. I could have chosen to focus on the city planning profession, in which case I would have emphasized the physical design elements of planning. Though such a focus might have pleased design-oriented practitioners, it surely would have alienated anyone interested in a more policy-oriented form of planning or in electric-power planning in particular. Happily, exemplars of such a story are already available; e.g., Scott 1969, Boyer 1983, Foglesong 1986, Hall 1988. Alternately, I could have focused on the linkage between planning and the social sciences, in which case I would dismiss physical design and focus on public policy making instead. Again, taking this tack might have pleased scholars and others who are interested in the design and execution of public policy, but it would lead many city planning practitioners to dismiss the book as being irrelevant to their needs and practice. Fortunately, an exemplar of this kind of history is also readily available: John Friedmann's magisterial *Planning in the Public Domain* (1987).[2] A third alternative would have been to focus exclusively on the historical role of planning within the electric power industry.[3] Though such a history would surely engage the interest of electric power practitioners, it would hold little appeal for the vast majority of both design- and policy-oriented planning scholars and practitioners.

I seek instead to construct a history that engages scientific planners and politicians, advocates and analysts, lay people and designers, and all other kinds of people who care about the ways in which cities and policies are planned and developed. I begin with the point of view provided by the urban and regional planning tradition (flawed though it is), but then show that tradition being transformed by its members, new as well as old, often in response to changing historical circumstances.[4] Consequently, the tale is one of continual dialogue between current and new members of the tradition, of continuing argument over the meaning of "planning," and of major substantive turns in planning's history.

This history begins, therefore, with a group of people (let us call them planners) who initially sought to solve the pressing problems of a newly industrializing and urbanizing society and to do so by guiding the growth and development of cities and regions. The first important turn occurred with the coming of the New Deal in the 1930s. At that time, a signficant number of planners began moving away from the physical design subtradition and toward a broader social-science-based perspective. With that turn, many planners began thinking of planning as a procedure for guiding decision making and action, not just for cities but for any enterprise, whether it be federal, state, or local, public or private. Many, perhaps most, design-oriented planners decried (and continue to decry) that turn.[5] This modernist effort to separate process from substance gradually led to a radical split between planning theorists and substantive practitioners, with the theorists engaging in esoteric arguments about what procedure should be followed when conducting any form of planning, and the practitioners (in transportation, in electric power, in housing) claiming to focus on the "real substance" of planning. They disagreed among themselves about what should be accepted as "planning" and who should rightly be called a "planner."

Though an important part of planning history, this modernist effort to separate process from substance was deeply misguided. In my view, the two are deeply intertwined, and the process by which planning is conducted depends in large part on the institutional context within which it takes place. Thus one should not be talking about the generic process of planning, but of planning rhetoric (persuasive argumentation) within specific institutions. The history of planning in America can, therefore, rightly be thought of as one of continuing fragmentation into diverse yet often overlapping rhetorics.

Electric power planning exemplifies the point. Electric power weaves in and out of the larger narrative of urban and regional planning in America, mediated primarily through the institution of regulated natural monopoly. Within that instititution, planners and managers and others form an interpretive community that speaks what I will call the normal discourse of electric power planning. Overwhelmingly dominated by privately owned firms that are regulated by state agencies, the electric power community speaks of electricity as a commodity, as an instrument for making profit, rather than as a public service (Rudolph and Ridley 1986; Nye 1990). Meanwhile, planners and elected officials at the local level tend to be scarcely aware of the electric power industry, thinking of it as just another business (though a rather large one).[6] That has not always been the case, however. At three critical junctures in the past—between 1898 and 1907, during the thirties, and in the early eighties—economic and political conditions compelled electric power planners and managers to engage in an abnormal discourse that extended well beyond the normal rhetoric of the institution. At those points in time, urban and regional planning was electrified, and the meaning of electricity and the future of the electric power industry were both hotly debated.

But I risk imposing too much too soon. Let me now present a more detailed history of planning in America. Then we will be in a better position to consider (in chapter 2) the merits of the rhetorical turn in general and the argument about planning as persuasive and constitutive storytelling in particular.

The Rise of Modernist Planning

Urban and regional planning has a long and rich history that is deeply rooted in both the physical design arts and the social sciences. The physical design roots can be seen in architectural treatises by Alberti, Palladio, and others; in utopian proposals (e.g., Thomas More's *Utopia*); in new towns (e.g., Charleville, France, in 1608); in plans for extending existing cities (e.g., Papal Rome in the fifteenth and sixteenth centuries); in garden and park designs (e.g., Andre Le Notre's design for Versailles); in Christopher Wren's plan for rebuilding London after the 1666 fire; and in L'Enfant's 1791 plan for Washington, D.C. (Lynch 1981; Barnett 1986).

But this planning was for the stable, hierarchical world of premodern, preindustrial societies. Throughout the later 1800s, Europe and

America industrialized and urbanized. Whereas towns and cities had previously grown slowly and organically in an almost evolutionary way, they now underwent sudden and radical transformation. Laissez-faire capitalism caused vast, historically unique cities to emerge where previously there had been only small tribal villages and natural ecosystems. Chicago, for example, grew from a population of 50 in 1830 to 300,000 in 1870 and 1,100,000 in 1890 (Mayer and Wade 1969; Keating 1988). People immigrated to these cities from the surrounding countryside and from foreign countries, such as Ireland, in search of a new and, they hoped, better life. Many of the immigrants went to work in the new industries that were forging the new urban and industrial order. Some brought with them new European ideas about how workers should relate to the men of capital; they and other workers organized to resist the more oppressive aspects of the new order. At the Haymarket Riot in 1886, at Homestead in 1892, at the Pullman strike of 1894, and at hundreds of other locations, they struggled for better working conditions. They also lived in oppressive conditions: the new cities were shockingly congested with people and traffic, their waters were badly polluted; communicable diseases (especially cholera) were common. Furthermore, most of these new immigrants were poor and lived in cramped, dark, dirty, and unhealthy buildings and neighborhoods (Riis 1971). Seeking to improve their lives, they often helped to elect and support political bosses and machines that distributed jobs and favors to friends. Thus the emergence of the modern industrialized city led to two partly overlapping impulses. The city created real and present problems for the people who lived and worked in them, but the city also instilled fear in the hearts of upper class elites. To them, the city had become a seedbed of labor conflict, political radicalism, and municipal corruption. Their social order was under siege.

Electric power had very little to do with the emergence and early growth of the modern industrial city. However, it did have a profound effect on the city's later growth, beginning with perfection of the electric streetcar system in the 1880s. Once perfected, this system rapidly displaced horsecars as the major form of urban transportation and thereby enabled urban development to extend well beyond the previous horsecar-based limits (Nye 1990). With new areas now opened to development, skilled workers could migrate out from the congested city center. Electric streetcars also simultaneously facilitated travel to downtown commercial centers and to new trolley parks at the outer ends of streetcar lines; city residents could travel quickly both to the department stores, hotels, and theaters in the city's center and to the fireworks, concerts, picnic grounds, and exciting rides of the amusement parks on its periphery. Privately owned, these electric streetcar systems became the objects of considerable political controversy. As David Nye puts it,

> Socialists and some progressives called for public ownership of all services, arguing that no private company should profit from a basic human need. Utilities . . . bribed politicians or contributed heavily to election campaigns in order to secure their charters and rights of way. . . . The streetcar lines thus became a center of class conflict, in which management could rally some support by appealing to prejudices against recent immigrants. (1990, 98)

So, conflict over the management of electric streetcar systems helped to focus public attention on how to guide change within newly industrializing cities.

How could the problems of the modern industrial city best be solved? Rejecting politicians as corrupt, and powerfully stimulated by Edward Bellamy's 1888 book, *Looking Backward,* many progressive reformers at the turn of the century began to think that rational design could produce rational cities. Some of them thought that for cities to thrive in the modern age, planners would have to creatively destroy the preindustrial city and build the modern city on its ruins. Baron Von Haussmann's effort to creatively destroy Paris for Emperor Napoleon III during the 1850s provided a powerful image of how such modern cities could be built (Pinkney 1958). Others (deeply influenced by Ebenezer Howard's 1898 book, *Garden Cities of To-morrow*) thought that people could be induced to move out of congested cities by building relatively small new towns ("garden cities") that would be dispersed thoughout the surrounding region, separated by greenbelts, and linked by planned canals and railroads (Bruder 1990). But the most important early image of how the problems of the modern industrial city could best be solved came in 1893 with the World's Columbian Exposition.

The Great White City and the City Beautiful Movement. The first major modern planning effort in America occurred at the 1893 World's Columbian Exposition in Chicago. Standing in stark contrast with its urban environs, the comprehensively planned and neoclassically designed landscape of this "Great White City" seemed to demonstrate quite powerfully that planning could bring both order and beauty to congested, chaotic cities.

But not just any order and any beauty; Daniel Burnham and the other men who built the Great White City believed that Truth and Beauty were timeless and universal and that those timeless principles could be

Burnham and his associates believed that the White City expressed timeless and universal principles of Truth and Beauty. When I listen to

Bach's Brandenberg Concerto no. 5 and scan photographs of the White City, I find something deeply appealing about that claim. But then I wonder whether "Negroes," women, working-class immigrants (like my mother's parents and grandparents), other contemporaries of Burnham, or Americans living today would agree. I find myself suspecting that the timeless and universal beauty of the White City actually expressed the beliefs and values of white men of capital in late nineteenth-century America. See Foglesong 1986 and Boyer 1983.*

used to beautify cities and thereby make them better places in which to live. "Let your watchword be order and your beacon beauty," Burnham said (Foglesong 1986, 124). Millions of people visited the Exposition and were awed by the vision of what a city could be, particularly when contrasted with the congested and chaotic ugliness of Chicago. Embedded in that contrast were two other significant differences that Burnham and his associates chose not to see: they designed the buildings of the White City so as to evoke the grandeur of past European civilizations, but they designed the spatial relationships between those buildings as if the past did not exist; and they proclaimed that White City beautification would make industrial cities better places to live, but they did nothing to provide decent homes and neighborhoods or provide safe and efficient means of transportation in and around the city.

Electricity played an extremely important role at the World's Columbian Exposition. Amply electrified, the Exposition helped to prefigure a future in which everything would be man-made and man would be "in total control" (Nye 1990, 34; see fig. 1.1). The exposition as a whole had more lighting than any city in the country at that time, and people moved around on an electrified sidewalk and railway and on elevators. Even the grand fountains were electrified. Replete with new electric appliances and manufacturing devices, the Electrical Building helped to display "man's increasing control over nature" and "the superiority of the present over the past" (41). It argued that advanced societies would be electrified, with electricity being strongly associated with Progress, Science, Light, Control, Order, and Democracy. Undeveloped societies would remain in primitive darkness. As Platt (1991) puts it, the Great White City "suggested a future world of efficiency, convenience, and leisure" (63), "promised an urban-industrial future worth striving for [, and] . . . helped make electricity the ultimate symbol of progress . . . a benevolent agent of moral uplift

*I have followed the language of the day in referring to African Americans, using, for example, "Negroes" till around 1966, then "Blacks," and finally "African Americans" in the 1980s.

FIG. 1.1 **The Great White City, Electrically Illuminated at Night**

and democratic values" (65). If electricity was the ultimate symbol of progress, then surely electric power planning would be the penultimate symbol, and professional planners would guide the electric transformation of society.

The electrified Great White City helped spark a "City Beautiful" movement that lasted for about twenty years. It did not take long, however, for observers to conclude that City Beautiful planning was not solving the most pressing problems of the time, particularly the traffic and population congestion problems that interested business elites of the major cities or the housing problems of immigrant populations. Criticized for not solving these problems, the City Beautiful soon transformed into the "City Practical" movement.

The City Practical. Proponents of the City Beautiful tended to think that the city was like a work of art and that the people who planned its development should be like artists. Alternately, proponents of the City Practical thought that the city was like a machine and that planners should be like the engineers or technicians who make machines run more efficiently. They believed that expert planners could—by collecting and analyzing data scientifically—plan a more efficient, convenient, and healthful system of land use and transportation in the city. As the planner-architect George B. Ford said at the time: if the task of planning is approached scientifically, then "one soon discovers that in almost every case there is one and only one, logical and convincing solution of the problems involved" (Foglesong 1986, 213).

The 1909 plan for Chicago was a key transitional event in the shift from beautifying cities to making them more efficient. Daniel Burnham, the lead planner for both the White City and the 1909 plan, inspired a generation of planners by telling them:

> Make no little plans; they have no magic to stir men's blood, and probably themselves will not be realized. Make big plans: aim high in hope and work, remembering that a noble, logical diagram, once recorded, will never die, but long after we are gone will be a living thing, asserting itself with ever growing insistency. (Wrigley 1983, 71)

Burnham and his fellow planners proposed broad new boulevards, a vast new civic center, an admirable series of parks along the lakefront, a new pier, and major changes in the rail and roadway system leading into and around Chicago (Moody 1911). Once again, however, this plan was not entirely consistent with the city that the private market economy was actually producing. In particular, the plan wanted the city's buildings to be of uniform height and for the civic center to be designed in neoclassical style, but Louis Sullivan and other Chicago architects were already designing and building a diverse array of new skyscrapers, varying both in height and design.

A key challenge for Burnham and the other planners in 1909 was to persuade the public to adopt and implement the Chicago plan, a plan that had been sponsored by an elite group of businessmen in the City's Commercial Club. "Above all things," one of Burnham's associates said, the city planner "must be able to show conclusively that the plan advocated is a plan not for a class but for all the people, and a plan not for a section, but for the entire city" (Foglesong 1986, 211). The Chicago planners' success in these efforts inspired others to believe that scientifically based physical planning could produce rationally and efficiently organized cities. But, equating the concerns of the business elite with the "public interest," these city practical planners ignored the cries for better housing and sanitation in the city's poorer quarters. And, persuaded of the truth and beauty of their plans, they finessed the political conflicts embedded in their elite view of the public interest. They created independent plan commissions that would mediate between expert planners, elected officials, and the public. "If the plan is really good it will commend itself to the progressive spirit of the times," Burnham said, "and sooner or later it will be carried out" (209–210).

The rational managers of emerging electric utilities also thought of their systems as machines, machines that were designed to transform natural resources into a commodity. At first the electric power "machine" was quite fragmented, consisting of many small, intercity lighting companies franchised by local municipalities. Gradually, however, system builders such as Samuel Insull of Chicago's Commonwealth Edison Company began buying out smaller companies (see Hughes 1983). That effort, when coupled with public hostility toward streetcar monopolies, led to local battles between the advocates of public and private power. Publicly owned systems made great headway between 1897 and 1907, thereby frightening private system builders into making a clever accommodation. Perhaps the key stroke in the transition was the system builders' argument that electric utilities were "natural monopolies" that functioned best as private corporations subject to regulation in the public interest by state commissions. Insull argued in 1898 that state regulation was desirable both to facilitate expansion of his company's system and to avoid municipal ownership. Decrying the evil of political machines, the National Civic Federation agreed in 1907, arguing that

> Public utilities are so constituted that it is impossible for them to be regulated by competition. Therefore, they must be controlled and regulated by the government; or they must be left to do as they please; or they must be operated by the public. There is no other choice. (Anderson 1981, 23)

As a result of these diverse reformist pressures, New York and Wisconsin became the first states to create public utility regulatory commissions in 1907. Twenty-seven other states followed their lead within the next six years. As Nye puts it,

> Electric companies were thus enshrined as "natural monopolies" to be regulated by state laws administered by commissions. This mixture . . . satisfied the trust-busting mood of the time, serving both as a form of progressive reform in which the private sector was formally controlled by the public, and as a kind of technocratic bureaucracy, in which decision making was taken away from the local communities and lodged in the hands of experts at the state level. . . . Utility issues were removed from the unpredictable local democratic process. (1990, 181)

From that time on (with the notable exceptions of the 1930s and 1980s), electricity was commonly defined to be a commodity produced by an increasingly large, complex, and privately owned machine. Given such a narrow interpretation, decision making in the electric power industry could be "relatively straightforward compared to what it would have been if engineers and managers had felt obliged—or indeed had been obligated—to fulfill social needs despite costs" (Hughes

1983, 464). Working for private natural monopolies that were subject to state regulation, and defining electricity as a commodity, electric power planners became disconnected from their urban planning counterparts. They talked about coordination, integration, stability, planning, order, control, and system and avoided talk about justice, fairness, and equity (Hughes 1983). Similarly disconnected from their electric power counterparts, subsequent urban planners who discussed infrastructure planning for the emerging "networked city" (Tarr and Dupuy 1988) almost always treated telephone, natural gas, and electric power systems as being factors beyond their control.

Regional Planning. During the 1920s, planners continued to use surveys and other forms of scientific research to find technical solutions to the practical, physical problems of the city. They further articulated the idea that cities could be planned comprehensively and that the city's land should be divided into discrete zones. Residences here, industries there, and commercial activities over there. Throughout the next few decades, planning consultants prepared hundreds of "comprehensive master plans" for cities. But as the twenties glittered on, many planners concluded that it would no longer be adequate to prepare plans for individual cities in isolation. Automobile traffic and electric power systems had grown enormously after World War I, thereby expanding metropolitan areas well beyond the boundaries of their core cities.[7] Accordingly, these urbanizing metropolises were better conceived as integrated regions that needed a regional planning perspective.

City Practical planning received its penultimate expression with preparation of the Regional Plan for New York and Its Environs during the 1920s. Funded by the Russell Sage Foundation, and well-connected to the business elite of New York City, this was the largest regional plan of its day and the first major plan to rely heavily on professional planners and social science data. Led primarily by Thomas Adams, a highly respected planner from England, these men sought to design a regional system of highways, airports, parks, and other facilities that would accommodate the population growth that they projected for the region (Johnson 1988).

Adams and the others were pleased with their work. And it had a profound influence on professional planners (Johnson 1988). But not all modernist planners likened the city to a machine or were so pleased with the New York Regional Plan. Lewis Mumford and other members of the Regional Planning Association of America (the RPAA), for example, thought that the region was like an organism and that planners should act like scientific gardeners or foresters (Sussman 1976). They sought to articulate an alternative view of development in which

Howard-esque garden cities would be dispersed throughout the region, connected with one another by "townless highways," adjacent to wilderness areas in which people could replenish their spirits, and generally be allowed to grow in a balanced and organic way. Mumford and the RPAA, therefore, explicitly criticized New York's regional planners for simply spreading the problems of the modern industrial city throughout the region. In the RPAA's view, planning should be an instrument of social change, and planners should plan a "fourth migration" of people out of modern cities and into "regional cities." That would require much greater reliance on the social sciences and a more radical politics of planning. As Mumford wrote,

> Genuine regional planning . . . is not content to accept any of the factors in city growth as outside human foresight and control. If we cannot create better urban conditions without changing our present methods and institutions and controls, we must be prepared to change them: to hold that the present means are sacred and untouchable is to succumb to a superstitious capitalistic taboo. (Scott 1969, 292)

Thomas Adams and the other New York regional planners found such comments to be hopelessly utopian. In Adams's view, Mumford and the others had a planning "religion" that was "unworkable." "Moreover," he wrote, "if planning were done in the way he [Mumford] conceives it should be done, it would require a despotic government to carry it out. I would rather have the evils that go with freedom," he wrote, "than have a perfect physical order achieved at the price of freedom" (292–293).

The City Practical's presumption that the city was like a machine reached its peak expression in the work of the French modernist architect Le Corbusier. In his 1922 design for Contemporary City (see fig. 1.2) and his later designs for Radiant City and for reconstructing downtown Paris, Corbu indicated that the city planner-architect should rely upon "pure reason" to impose the modern spirit of exactitude and order on cities. Accordingly, the modern planner should design entirely new cities or eradicate vast portions of existing cities in order to make them anew. History should be ignored, eradicated, or memorialized in special museums. In Le Corbusier's Contemporary City, people would work in sleek new downtown office towers and travel to them on high-speed highways (Fishman 1977).

The potential benefits of electrification formed a crucial element in the debate over regional planning. Seeking to diversify their customer base (which they called "load") and hence use capital more efficiently, and interpreting electricity simply as an economic good, the private

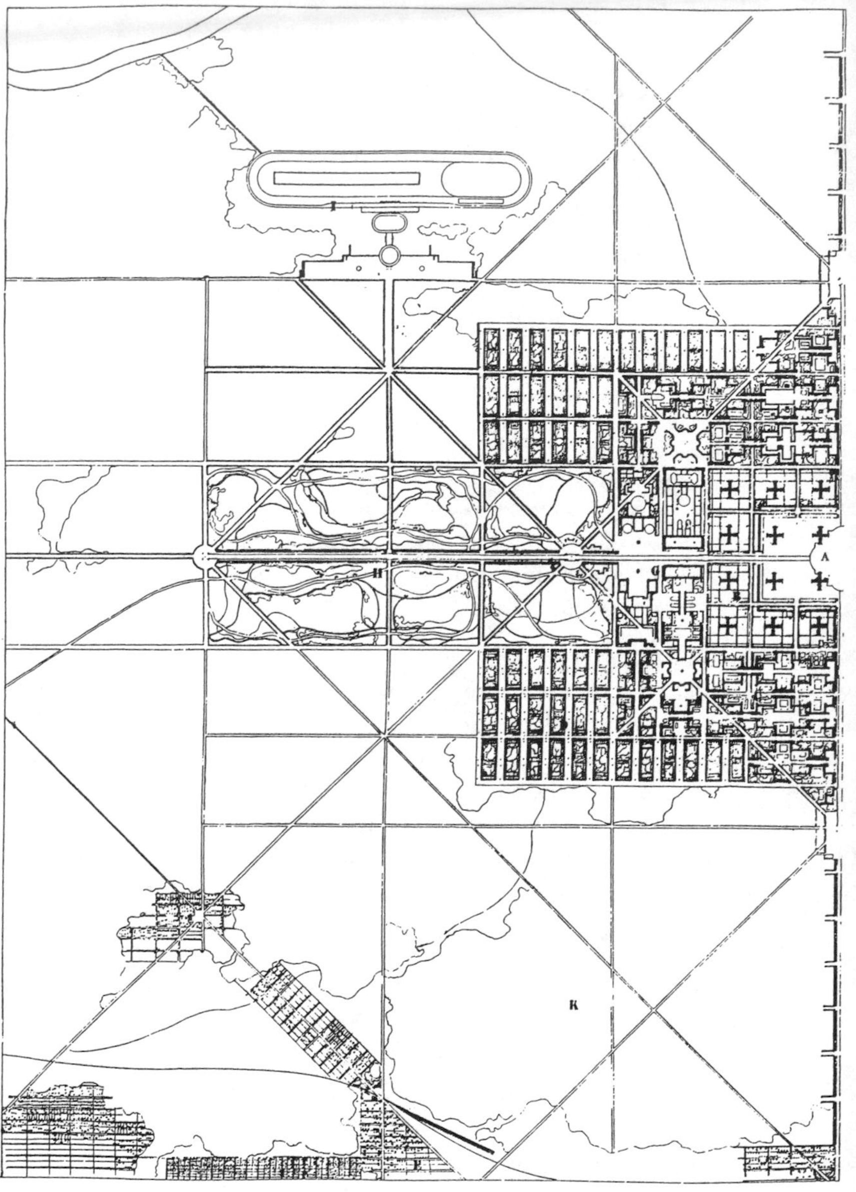

A CONTEMPORARY CITY

The heavy black lines represent the areas built upon. Everything else is either streets or open spaces. Strictly speaking the city is an immense park. Its lay-out furnishes a multitude of architectural aspects of infinitely varying forms. If the reader, for instance, follows out a given route on this map he will be astonished by the variety he encounters. Yet distances are shorter than in the cities of to-day, for there is a greater density of population.

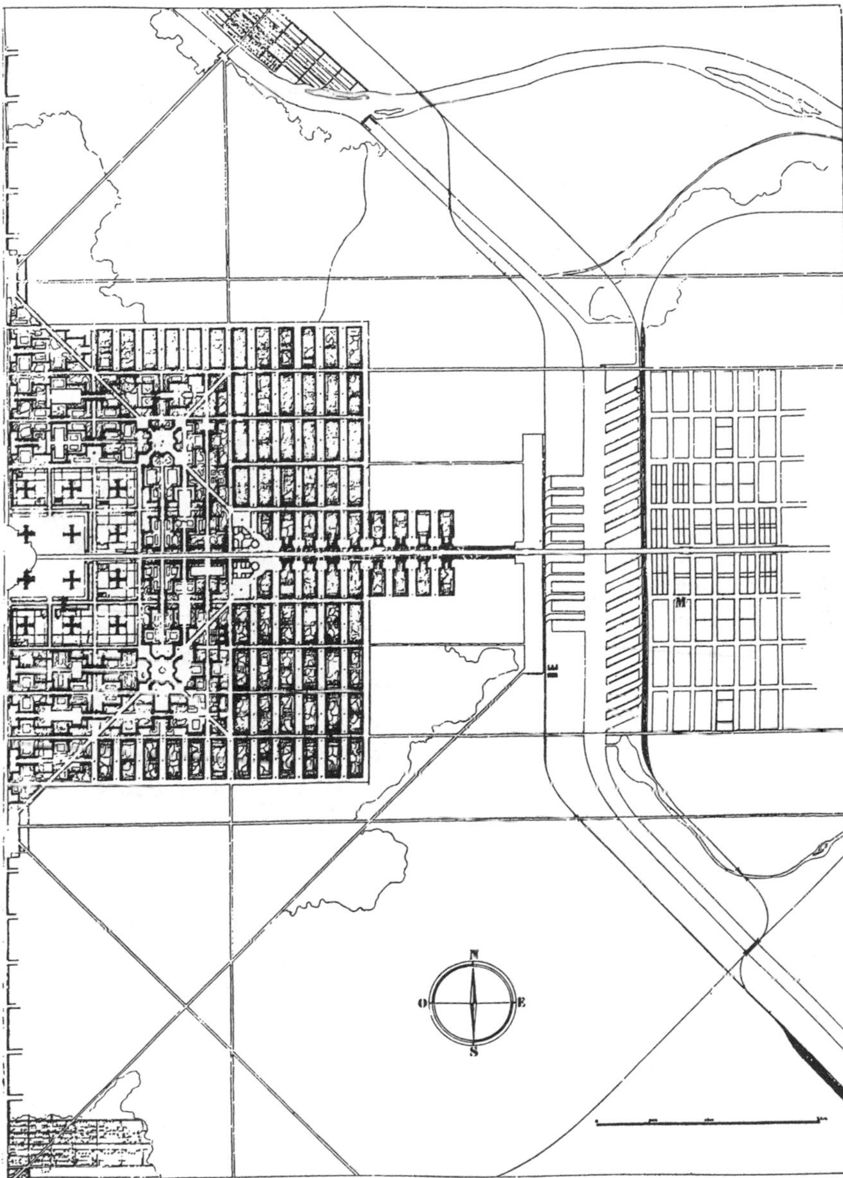

A. *Station.*
B. *Sky-scraper.*
C. *Housing blocks with "set-backs."*
D. *Housing blocks on the "cellular" system.*
E. *Garden cities.*
G. *Public Services.*
H. *Park.*
I. *Sports.*
K. *Protected zone.*
M. *Warehouses, Industrial city, Goods station.*

FIG. 1.2 **"A machine to live in": Le Corbusier's Plan of Contemporary City, 1922**

system builders had ample reason to extend their electric-power systems further into the developing fringe of urban areas, both following and facilitating the process of suburbanization. This process had at least three effects of great subsequent importance for planning.

Comfortably isolated in the control rooms of such regional electric power systems, electric power planners could "rationally" guide system change. Even as they did so, however, rational electric power planners were helping to create regions that were more socially segregated and politically fragmented, and hence less consistent with the underlying assumptions of scientific planning.

Perhaps the most important long-term effect was that regional expansion of electric power systems helped disperse the city spatially while also binding its economy more tightly to a centralized administration (Platt 1991; Nye 1990). Regional electrification helped people sort themselves into diverse groups by physically moving into separate, homogenous enclaves spread out over an ever-larger region. Though separated, people remained customers of tightly integrated and centrally managed regional systems. Thus regional electrification helped to promote "not only urban deconcentration and social segregation but cultural and economic centralization" (Platt 1991, 285).

Regional electrification also had ecological effects that proved difficult to see and hence grew over time. Progressive reformers and feminists of the time spoke of electric power as a "drudge and a willing slave" (i.e., a natural resource) that would free housewives from housework (Nye 1990). Though electrification did make individual household tasks easier, use of the "slave" also connoted a radical separation of the ecological place of electricity production from the economic point of consumption, and thus made it harder for people to be aware of (or truly account for) the social and environmental costs and consequences of using electricity. Energy flows became less and less visible to users of the new electric appliances: they could not see the coal being mined, the dams being built, the oil being pumped, or the pollution being generated as a direct consequence of appliance use (Cronon 1991; Stern and Aronson 1984).

Closely linked to the creation of vast, privately owned regional systems guided by profit was the deterioration of living conditions in inner city slums and the rural countryside. By the end of the twenties, most older cities had bifurcated into new electricity-intensive areas and older premodern neighborhoods, wherein obsolete and decaying housing coupled with overcrowding and exploitation by landlords to produce appalling living conditions (Platt 1991). Electrification was slow to reach farmers as well. Extension of distribution lines into the country was costly and the farm population was declining, so private systems had little economic motive to supply farms with electric power. The end result was that by the late 1920s only about 10 percent of all farms had electricity.

The failure of private systems to supply the urban poor and rural farmers with affordable electric power led to a policy debate over the future of regional systems. As electric power systems became more regional in scope during the 1920s, private utilities began writing and talking about "superpower" systems. Such regional systems would be interconnected with one another, controlled by a small number of large corporations, and regulated by regionally harmonized state commissions. Superpower would, W. S. Murray (1925) argued, allow the nation to grow without falling victim to the seductive appeal of communistic propaganda. Contrary to Murray, many others (such as Gifford Pinchot, then governor of Pennsylvania) extolled the virtues of publicly owned utilities and decried the inequities imposed by privately owned ones. To them, public ownership of utilities was part of the Progressive-Era move toward conservation and reform. In 1924, Joseph K. Hart (a member of the RPAA) wrote that "giant power, under public control, with power distributed to all on equal terms, offers economic freedom to humanity, the hope of communities within which intellectual freedom can be realized and the culture of spirit will be posssible" (Funigello 1973, 35).

National Planning and the Coming of Process Theories of Planning. By the end of the 1920s, a few planners had begun to look for an alternative conception of planning. Best personified by Mumford and others in the RPAA, these innovative critics argued that city practical planners simply tinkered with growth at the margin. In their view, mainstream planners could not explain how and why cities and their metropolitan areas grew, were not prepared to use knowledge about those processes to guide development, and did nothing to solve the housing and social problems of older cities or to bring farms into the modern era. A broader-scale form of planning, one more deeply rooted in the social sciences, was needed.

The Depression brought these issues to the fore and led to further changes in urban and regional planning. Most importantly, the Depression caused most urban planners to conclude that the private market should not be left alone to guide the nation's (hence the cities') development, and that effective planning required coordinated action at all levels of government (Hancock 1988). Many planners began to think of planning as a way to solve a wide variety of public problems rather than as the preparation of end-state physical plans for a city's future development. A process rather than a physical design.

Excitement ran high among reform-minded planners. New Deal planners contributed their skills to temporary work relief efforts, major public works projects, soil conservation, industrial recovery, slum clearance and housing programs, population resettlement through Greenbelt Towns, regional planning with the Tennessee Valley Authority (TVA), and national planning with the National Resources Planning Board (NRPB). Perhaps now they could design cities, regions, indeed the nation as a whole, in a more rational and coordinated way.

The NRPB lasted from 1933 to 1943. During that time, it had modest success in promoting planning for public works projects and in coordinating federal planning activities, and it had greater success in stimulating city, state, and regional planning and in gathering data and sponsoring influential research projects (Hancock 1988; Clawson 1981). Perhaps most importantly for the subsequent development of planning, the Urbanism Committee of the NRPB (1937) produced the first major study of American cities in a national context. The Urbanism Committee also redefined planning as a permanent administrative activity involving *rational procedures of choice,* an activity that was closely connected to elected officials yet still focused on the physical form and development of cities and regions. Lacking structural authority, short on funds and political support, and worn down by accusations of promoting socialism, the NRPB ceased to exist in 1943.

The New Deal also gave planners a new opportunity to cope with worsening housing conditions. A study by the Civil Works Administration during 1933 and 1934 revealed a widespread pattern of slums and poverty in urban America. In response, the Roosevelt Administration sought to replace slums with public housing rental units for low- to moderate-income groups, and to subsidize construction of single-family (mostly suburban) homes for middle- to upper-income white families. By 1937, the Public Works Administration had built over 20,000 rental units, all segregated by race. In that same year, Congress enacted new housing legislation (the Wagner Act) which shifted priority away from providing rental public housing and toward clearing slums and rehousing the poor in the cleared areas. The Roosevelt Administration also created a Federal Housing Administration that insured long-term loans for new single-family homes. Though successful on their own terms, the slum clearance, public housing, and home loan programs sped up the process of inner-city decay and accelerated the process of racial and income segregation in metropolitan areas (Hancock 1988).

The TVA proved to be at least as influential as the NRPB, and far more durable (Creese 1990). Deeply influenced by the RPAA's conception of regional planning, indeed by the argument for "giant power" that Pinchot and others had promoted, the TVA was created in 1933. Covering the 41,000 square mile Tennessee River Basin, TVA became the single most important example of regional planning implementation by the federal government. It sought to combine navigation improvements, flood control, soil erosion control, construction of hydroelectric dams, agricultural improvements, and new town construction into a single coordinated effort to bring an entire region out of poverty. Though TVA proved quite successful at building dams (and later at building coal plants, strip mines, and nuclear plants), it proved far less successful in its regional planning efforts. Bureaucratic infighting, vague enabling legislation, inadequacies in regional planning theory, continuing accusations of socialism, and new priorities established by the coming of World War II and the Cold War rather quickly eroded TVA's regional planning effort. Indeed, by 1936 TVA had essentially become a public power utility. In the end, the TVA experience taught planners the limits of comprehensive regional planning in America.

TVA would not have been created had it not been for widespread public antipathy toward private business in the early 1930s. It was part of a broad effort to reform the electric utility industry and to distribute the benefits of electric power more fairly across the spectrum of American society. Congress enacted the Public Utility Holding Company Act in 1935 and the Rural Electrification Act in 1936, and local elected officials tried to municipalize power systems. The first act gave the Securities and Exhange Commission the power to dissolve any holding company (a tiered grouping of privately owned utilities) that could not demonstrate its usefulness, whereas the second created a Rural Electrification Administration which brought power to rural areas that private utilities had refused to serve. It also instructed a revamped Federal Power Commission to ensure that the nation had an adequate supply of electric power at rates that would be "just and reasonable" to the consumer. The New Deal also created federal power marketing agencies (e.g., the Bonneville Power Administration) to tap hydroelectric resources and market power at low rates, with first priority going to publicly owned utilities. For a moment, it appeared as though electricity would be defined as a public service rather than merely as a commodity.

These challenges notwithstanding, privately owned electrical utili-

Corn and Horrigan (1984, 49–50) describe the effect of Futurama as follows: "The darkened space, the deep comfortable chairs, the soothing but authoritative voice of the narrator, the minute realism of the vast model—what an atmosphere for persuasion! Certainly no futuristic film or exhibit had ever been so convincing." For reasons discussed in chapter 2, I would characterize Futurama as an extremely powerful trope in the modernist planners' story of control.

ties and associated businesses thrived throughout the Depression. They did so in part by reassuring the public that electricity would light the way to an ideal technological future. This effort to reassure the public culminated in the New York World's Fair of 1939 (see Corn and Horrigan 1984; Kihlstedt 1986; Nye 1990). Visiting the "Futurama," "Democracity," and "City of Light" exhibits, fairgoers were presented with extraordinarily optimistic images of the world of tomorrow, images that were powerful and compelling and which made the modern city of the future appear to be both desirable and inevitable. Riding in comfortable moving chairs above a scale model of the ideal "City of 1960," floating overhead as if in a low flying aircraft, visitors to General Motors' Futurama (designed by Norman Bel Geddes) watched and listened in awe as a narrator expounded the wonders of the world of tomorrow. And inside the Perisphere at the Fair, people looked down on Democracity, a vast diorama of the city of 2039. There they saw an ideal commercial center surrounded by satellite suburbs and, further out, by tilled and grazed agricultural land. Much like Futurama, Democracity had a powerful, if selective, effect on visitors. As Nye puts it:

> Using lighting, music, film, and scale modeling, Democracity was a theatrical performance that achieved a powerful illusion. But it omitted a great deal. There were no slums inhabited by ethnic minorities, or poor neighborhoods or run-down single-family homes. There were no traffic jams, no unsightly factories, no unemployment, no polluted streams, no smog, no industrial blight. It contained no suggestion of a large standing army or advanced weaponry. In short, science and technology had no ill effects in this utopia. (1990, 371–72)

And in Consolidated Edison's City of Light, which provided a scale model of the entire New York metropolitan area, visitors were guided not by a human narrator but by a voice from the sky that presented a dramatic word picture of the ever changing city. The voice of the modernist planner presented the modernist city as if it were a historical actuality, a city powered by electricity that soon would be too cheap to meter.

In sum, planning became an integral part of American society during the 1930s, but it also left an ambiguous legacy. New Deal planners successfully used their skills to address a wide variety of problems, but their effort and new initiatives were widely condemned for being socialistic. The NRPB redefined planning as a process that should include concern for low-income people and resource redistribution but that left urban planners divided about whether they should accept such a broadened definition or continue their past focus on the physical design of the city. Perhaps most important, the New Deal era ended with most planners being eager to define their practice in apolitical terms, terms that would enable planners to apply their new skills and expertise without incurring the wrath of conservative businessmen. In practice, that meant limiting the attention of planners to compensating for market failures, planning major public infrastructure investments, and ameliorating negative market externalities. As urban planners narrowed their focus, the private market continued to develop and refine new technologies (most notably the automobile, the airplane, electric power, radio, and television) that greatly facilitated the movement of people and manufacturing plants out of central cities and into the suburban fringe.

"Rational Planning" after World War II: Apolitical Guidance from the Central Plateau? In the decade following the end of World War II, urban and regional planners began to focus their attention on two problems (or opportunities): explosive growth in the Sun Belt and in the suburban peripheries of major cities, and stark deterioration of inner cities. That pattern of growth and decay resulted in part from the way private investors chose to invest and disinvest. But it also flowed from patterns of racial bias and segregation deeply ingrained in American society and from new federal legislation that reinforced and facilitated those patterns with regard to market investment. Tract houses mushroomed at the outer edges of American cities throughout the 1950s, while inner city neighborhoods deteriorated and lower-income "Negro" immigrants from the south moved into new public housing projects in the big cities of the Northeast and Midwest (Lemann 1991).

Responding to inner-city "decay," Congress enacted the Housing Acts of 1949 and 1954 and thereby encouraged planners to "redevelop" and "renew" urban areas. Robert Moses, who guided public investment policy in the New York area from the 1920s through the 1960s, and others tore down old neighborhoods and replaced them with high-rise low-income housing, commercial/office centers, and interstate highways. He and other public devel-

Robert Moses often said that "You can't make an omelet without breaking eggs" and "If the end doesn't justify the means, what does?" (Caro 1974, 218). He meant that

planners would have to hurt some people in order to build the highways, urban renewal projects, parks, dams, and other major public structures required by a great city. "Whose omelet?" one might ask. In his later years, Moses became a power broker who was more interested in accumulating power than in helping people construct a good city and a good region (Caro 1974).

opment planners transformed public housing projects into clusters of high-rise structures not unlike those that Le Corbusier had envisioned in the 1920s (Hirsch 1983; Bauman 1987). Moses, widely known as the man who "knew how to get things done," and others creatively destroyed the inner parts of their cities in order to make them new (Berman 1982).

Construction of the interstate highway system pursuant to the 1956 Federal-Aid Highway Act greatly facilitated migration of people and industry to the Sun Belt, travel to and from the burgeoning suburbs, and the effort to renew inner cities (Rose 1990). More and more people drove more and more miles consuming more and more gasoline on these new highways.

Urban renewal and interstate highway planners were heavily influenced by a new process theory of planning. Led by "Chicago school" theorists in the early 1950s, planners began to think in terms of *rational planning* and to rely heavily on systems analysis and computers. Thinking of planners as "generalists with a specialty," these theorists initially redefined planning as a separate academic discipline, as "really reflective decision making" best conducted by central experts.[8] But gradually they redefined planning to mean (1) a generic process that applied to more activities than just physical planning; (2) a decision making process rooted in the social sciences (rather than an architecturally based design process); (3) an activity that focused on public policy (rather than on cities as such). In this context, Edward Banfield introduced the "rational planning" model into the planning literature. In an appendix to Meyerson and Banfield's (1955) classic *Politics, Planning, and the Public Interest,* Banfield defined planning as the "rational selection" of "a course of action to achieve ends," wherein a "rational" decision maker (1) "considers all of the alternatives (courses of action) open to him"; (2) "identifies and evaluates all of the consequences which would follow from the adoption of each alternative"; and (3) "selects that alternative the probable consequences of which would be preferable in terms of his most valued

Notice that "rational planning" takes place inside one man's (the "decision maker's") head and that the decision maker's goals are defined in advance. What happens when the planner or decision maker discusses his goals and plan with others, especially when those others have differing goals and concerns?

ends" (314). Banfield also sought to define other key terms clearly and precisely. Thus he defined a *plan* to mean "a course of action which can be carried into effect, which can be expected to lead to the attainment of the ends sought, and which someone . . . intends to carry into effect" (312); a *course of action* to mean "a sequence of prospective acts which are oriented toward the achievement of ends" (312); and an *end* to mean "an image of a state of affairs which is the object or goal of activity" (304). Banfield and other rational planners paid little attention to what those goals might be. Presuming that the end of ideology had come and that pluralist politics provided all interested parties with an equal chance to influence goal selection, they told planners to attend to means and to leave goal selection to others.

Two years later, in a seminal statement about planning education, Harvey Perloff (1957) argued that these new rational planners should be trained to be "generalists with a specialty"; they should be taught core principles that apply to all planning activities, and they should be taught the detailed, substantive facts and concepts of particular areas of policy concentration. Electric power planning, for example. Or transportation planning. By the late 1950s and early 1960s, the rational planning paradigm had largely taken over the profession (Perin 1967). As the urban geographer David Harvey (1989) puts it, such planners sought to "master" the metropolis as a "totality." They produced "large-scale, metropolitan-wide, technologically rational and efficient urban *plans,* backed by absolutely no-frills architecture" (p. 66). The metropolitan transportation planning process first used in the Chicago Area Transportation Study of 1955–1962, then later incorporated into the 1962 Federal-Aid Highway Act, provides a clear example of rational planning as applied in practice (see Black 1990).

The rational planning model guided the expansion of electric power systems as well, though urban planners know little about that experience. The industry had emerged from World War II with a faith in what appeared to be a limitless future. Stimulated in part by the industry's rate structure (which charged less per unit for each additional unit of electricity demanded) and all-electric home advertising, demand for electricity grew dramatically: the industry's total national power output surged from 307 billion kilowatt-hours (kWh) in 1947 to 2,003 billion kWh in 1975. Total generating capacity grew almost as dramatically, leaping from roughly 65,100 megawatts (MW) in 1947 to 527,600 MW in 1975 (Loftness 1978, 167, 169). The newer generating units grew in size as well, increasing from a maximum of 200 MW in 1940 to a maximum of 1,000–1,200 MW in 1975 (U.S. Department of Energy 1981). The cost of power generation continued to decline, which in turn stimulated demand. In this context, rational planning in the electric power industry focused on identifying the least costly way of generating power and on finding the least-costly locations for new power plants and transmission lines.

At first, the expansion of power generation capacity was limited to plants fired by fossil fuels. However, the industry began giving serious consideration to nuclear energy as a potential source of electric power in the mid-1950s. From the time the first nuclear power plant went into operation at Shippingport, Pennsylvania, in 1957, to the time that Jersey Central Power and Light purchased a "turnkey" reactor in 1963, the utilities' general faith in Progress was largely replaced by a very specific faith in nuclear power. From five orders for 776 MW of reactor capacity in 1955, demand for nuclear power leaped to thirty new orders for 25,633 MW in 1967 (Goodwin et al. 1981). Without benefit of any independent analysis of the actual cost experience of the new technology, electric utilities enthusiastically embraced what Bupp and Derian (1981) call the "great bandwagon market" for nuclear power in the mid- to late sixties.[9]

Widely used, the rational planning model soon encountered problems. It assumed that planners could specify goals that perfectly represented society's values, use lawlike scientific generalizations and associated predictions to choose the one best way of achieving those goals, and count on some central authority to implement the preferred alternative. None of those assumptions proved valid in the America of the 1960s, though the structure and normal discourse of particular institutions—most notably regulated natural monopolies—made it possible for some planners to believe that the assumptions had merit. For a brief moment, many planners were torn between the rational planning ideal and the argument for "incrementalism" embedded in Charles Lindblom's devastating 1959 critique of the rational model. Public administrators do not decide in accord with the rational model, Lindblom argued; the method is simply impossible to apply to complicated problems. Administrators set a simple goal (thereby ignoring many values and possible consequences), outline a few policy alternatives, compare those alternatives in terms of past experience with small policy steps, then choose an alternative by simultaneously considering ends and means. "Policy-making," in Lindblom's view, was "a process of successive approximation to some desired objectives in which what is desired itself continues to change under reconsideration" (164–65).

By the end of the 1950s, scientific planners thought that experts could "rationally" guide change, whereas political planners thought that politicians and administrators guided change by taking small, incremental, and politically acceptable steps away from the status quo. Events of the 1960s upset both views.

Reactions against Modernist Planning

A contemporary literary critic, Marshall Berman (1982), suggests that building the modern city can be a creative venture, but that living in it can be a nightmare. Guided by the rational planning model, and confident in the validity of that process, urban renewal and regional transportation planners paid little attention to the adverse effects of their plans on lower income people. If the alternative was already the "best," then any harm to those people would have already been incorporated into the planners' net benefit calculation. All around the country, urban renewal and regional transportation planners—working in symbiotic interaction with market investment patterns (including those of electric power utilities)—destroyed diverse, vital, high-density urban neighborhoods. What is more, urban renewal, freeways, and investment patterns selectively destroyed neighborhoods occupied primarily by "Negroes."

This pattern of selective "renewal" stimulated new voices to speak (and be heard) in planning. Shocked by the destruction of those neighborhoods, Jane Jacobs attacked planners and celebrated urban diversity and vitality in her influential 1961 book, *The Death and Life of Great American Cities.* Herbert Gans (1962) reinforced Jacobs's attack by documenting how urban renewal had devastated a vital ethnic neighborhood in Boston. And Michael Harrington (1962) drew attention to the plight of the poor in an affluent America. From their point of view (see fig. 1.3), the city looked quite different than it did from the modernist planners' gleaming white towers or from the chairs floating above Futurama.

But these early voices barely scratched the surface of tensions that were teeming in the dark underside of modern America and its affluent cities. When Rosa Parks refused to move to the back of a Montgomery, Alabama, bus in 1955; when four freshmen students sat at a whites-only lunch counter in Greensboro, North Carolina, in 1960; when thirteen men and women began the first Freedom Ride in May of 1961; when demonstrators braved fire hoses and snarling dogs in the Birmingham marches of 1963; when thousands of people marched in Washington, D.C., in August of 1963; when Negroes and whites protested in hundreds of locations throughout the country in the name of freedom and civil rights for Negroes, urban and regional planning began to change quickly, dramatically, and (perhaps) irreversibly (Garrow 1986).

FIG. 1.3 **John Sloan's *The City from Greenwich Village*, 1922**

In 1964, President Johnson launched a "War on Poverty" in pursuit of his "Great Society." He and Congress created Medicare, Medicaid, the Job Corps, the Head Start, and other programs in quick succession. But perhaps the most important initiative from a planning point of view was enactment of the Economic Opportunity Act of 1964, the subsequent creation of Community Action Programs throughout the nation, and the effort to facilitate the "maximum feasible participation" of low-income Negroes and whites in those programs (Marris and Rein, 1967, 1982).

Affected by Harrington and other influential authors, inspired by the civil rights movement, rejecting the consensual and authoritative assumptions of rational planning, and motivated by President Johnson's social planning initiatives, many planners sought to assure that the interests of ethnic minorities and lower-income people were represented in planning processes. In 1965, Paul Davidoff argued for *advocacy planning*. "Appropriate planning action cannot be prescribed from a position of value neutrality," he wrote, "for prescriptions are based on desired objectives" (Davidoff 1965, 331). Planning is an adversary process, he argued; "the planner . . . should be an advocate for what he deems proper [and] . . . represent and plead the plans of many interest groups" (331–32). Note, however, that Davidoff did not reject rational planning; rather he thought that advocate planners could make planning more rational and more just by better informing the public of available alternatives and by forcing those who were critical of establishment plans to produce superior ones.

But events quickly outpaced Davidoff's advocacy model. Blacks living in the big cities of the northeast and midwest (not to mention Los Angeles) felt that progress was not being made quickly enough. Cities erupted in flames. First came the Watts neighborhood of Los Angeles in 1965, Newark and Detroit in 1967, and then many other cities. The sudden appearance of the war in Vietnam greatly exacerbated those tensions. Just three months after being reelected, President Johnson began bombing North Vietnam. American troops poured into the country (by June 1967 there were 448,000 there), and students began protesting then resisting his efforts to draft them into the war. Like many other Americans, some planners concluded that mere advocacy was hopelessly insufficient. Seeking to respond to the crisis in the cities while also reining in unruly participatory-based community action programs, the Johnson administration enacted the Model Cities Act of 1966. This Act sought to coordinate action throughout metropolitan areas in such a way as to aid the poorest, most deteriorated city neighborhoods (Fox 1986). Too little, too late?

The very idea of rationally planning and coordinating action in the public interest seemed to evaporate in the "violence shock" of 1968 (Gitlin 1987). In January, the North Vietnamese and Viet Cong launched the Tet Offensive, thereby demonstrating that Americans were not winning the war. Two months later, President Johnson declined to run for reelection. Five days after that, Martin Luther King, Jr., was assassinated and Blacks ravaged cities in a vast outpouring of anger and grief. In June, Robert F. Kennedy was assassinated. Two months later, Lincoln and Grant Parks on the lakefront in Chicago became the site of a violent police riot against student protesters who sought to awaken what they called the "sleeping dogs" of the Right (Gitlin 1987).

Appalled by the more radical elements of the antiwar movement, and frightened by the flames of Watts and Detroit, a majority of the American electorate backed off from Johnson's liberal reform efforts. With President Nixon's election in 1968, the federal government began a two-decade-long conservative march away from inner cities and social programs and, though not completely abandoned, planning came to be associated with the "failed programs" of the sixties. Urban planners themselves left the decade split between those who focused on physical planning and those who focused on social planning, between rational planners and advocate and radical planners.[10]

The civil rights movement expressed the idea that oppressed people should fight for themselves, that they cannot and should not count on someone else to represent their interests. That idea spread throughout the late sixties and early seventies, most notably into the environmental and women's movements. Rachel Carson's tremendously influential 1962 book, *Silent Spring,* had heralded the arrival of the natural environment as yet another voice in planning. Then Congress enacted environmental legislation throughout the 1970s that required detailed planning for air quality, water quality, hazardous waste disposal, toxic materials handling, and for assessing the environmental impacts of major federal actions (Hays 1987; Gottlieb 1993). Many planners took positions with federal and state environmental agencies, and many others found themselves wondering whether there were "natural limits to economic growth" (Meadows et al. 1972) and trying to manage growth in Florida, California, and other parts of the country.

The environmental movement imposed several new requirements on electric utilities. Beginning with the National Environmental Policy Act of 1969 and the Clean Air Act of 1970, they were required to install expensive new electrostatic precipitators and scrubbers to cleanse their stack gas emissions of air pollutants such as particulate matter and sulfur dioxide (see Roberts and Bluhm 1981). They were required to install expensive new cooling towers to avoid damage to receiving waters. Closely monitored and frequently taken to court by advocates working for environmental organizations, utilities found their newest power plants becoming more expensive to build and more difficult and time-consuming to site.

But just as the civil rights and antiwar movements radicalized planning as a whole, so too did they help to radicalize planning and action in the electric-power industry, particularly with regard to the siting and construction of new nuclear power plants. Nuclear power had grown rapidly: from ten commercially licensed reactors with an installed capacity of 933 MW in 1965 to 554 plants with a capacity of 38,568 MW in 1975 (LeBel 1982). At first in the courts and then through marches, blockades, referenda, and other forms of nonviolent protest, opponents tried to stop any further expansion of nuclear power. Emphasizing the connection between nuclear power and nuclear war, and relying on grassroots organizing, antinuclear activists created organizations such as the Clamshell Alliance and the Abalone Alliance and tried to block new nuclear facilities at Seabrook, New

Hampshire, and Diablo Canyon in California (see Rudolph and Ridley 1986).

The feminist movement brought yet another voice to planning during the 1970s. Urban spaces and decision processes were "gendered," they argued, much to the harm of women (Hayden 1984; Milroy 1991). Tending to bear primary responsibility for child rearing and for housekeeping, while also working full- or part-time away from the home, many women discovered that they had transportation needs that were going unmet. Furthermore, many single women and women who headed families often found that suburban zoning ordinances made it difficult for them to obtain adequate housing (Abu-Lughod 1991).

By the mid–1970s, modernist planning no longer ruled supreme. Though still thriving in narrow institutional arenas such as the electric power industry, modernist planning and its proclaimed ability to discover the Truth through scientific research and to perfectly represent the public interest in political arenas was fast becoming just one of many voices in planning.

Planning in a Post-Modernist Era

Like 1968, 1973 proved to be a pivotal year. In that year the Vietnam War came to an end, at least for Americans on the field, and the unifying object of political radicalism disappeared. Almost simultaneously the great economic boom of 1945–73 came to an end, for in that year the Organization of Petroleum Exporting Countries (OPEC) oil embargo touched off (or dramatically accelerated) a series of economic shocks that soon came to be known as "stagflation" (a combination of inflation and recession) and "deindustrialization" (Bluestone and Harrison 1982).

The most immediate effect of the oil embargo on electric utilities was to increase the cost of fuel oil. Utilities sought to pass those costs along to consumers as quickly as possible, which in turn stimulated consumers to oppose rate increases. Outraged at the way rising costs were simply passed along to them, and not particularly concerned about the utilities' financial woes, newly formed consumer groups pressed for major changes in the way that rate increases were allocated (Gormley, 1983; Rudolph and Ridley 1986). The oil embargo also led to the creation or growth of numerous alternative energy groups, community economic development initiatives, and "appropriate technology" grassroots movements. Partly spawned by the antiwar and counterculture movements of the late sixties, and searching for environmentally benign ways of increasing local self-reliance and economic development, these groups found much theoretical support in several publications of the mid- to late 1970s; e.g., Schumacher 1973, Lovins 1977, and Morris 1982.

As the economic effects of the OPEC oil embargo rippled through the nation's economy, whole regions went into economic decline. Dismayed by the seeming collapse of the idea of Progress and the American Dream, many people concluded that modernist planning could not deliver on its promise to guide Americans into a bright new tomorrow.[11] Most Americans, planners included, got down to business. Many found themselves charged with preparing local economic development plans, plans that sought to enhance their cities' abilities to compete in an increasingly globalized economy.

But as they did so, even businesslike planners found their work widely discredited. With the election of Ronald Reagan as president in 1980, one commonly heard conservative elected officials blaming the country's economic decline on "big government," on the "excesses" of Johnson's Great Society initiatives, on "failed governmental programs," and (by association) on planning. Blaming public planners, arguing that markets should be deregulated, insisting that economic development should be guided by the "competitive marketplace," and reflecting the interests of middle-class suburban constituents who had been frightened by the excesses of the civil rights and antiwar movements, the Reagan administration effectively abandoned cities and their problems.[12] If jobs and people increasingly moved to the "exurban" fringe as a result (Garreau 1991), if the economic plight of the people left behind (who were far more likely to be African-American or Hispanic) worsened, then so be it. Declining cities—with their increasingly low-income and minority populations and their African-American Democratic mayors—would have to fend for themselves.

Confronted with this retreat from the ambitions of modernist planning, planning scholars strove to find a replacement for the rational model that they and others had found so wanting in the sixties (Galloway and Mahayni 1977; Hudson 1979; Burchell and Hughes 1980; Hemmens 1980; Alexander 1984). Some sought to synthesize rational planning and incrementalism. Calling the result *policy analysis,* they sought to make public policy decisions more rational by providing expert advice to powerful decision makers (or clients) about specific programs (Quade 1975; Wildavsky 1987; Alterman and MacRae 1983; McClendon and Quay 1988). Other planning theorists rejected the dominant notion of planning as a rational, technical activity either in its pure form or in its policy analytic or advocate planner forms. If people disagreed about fundamental goals and values, they argued, then rational planning and its variants are just ways to impose unwanted goals and criteria on other people. In their view, planning theorists should study planning practice and find ways to open that practice up to diverse voices (Forester 1980; Forester 1982; Schon 1983; Baum 1983; Hoch 1984; Burchell 1988; Hoch 1990). But, as the city seemed to fragment and dissipate and diverse groups seemed to construct diverse understandings of "the city" and of "planning," it became less and less clear who should be included as a practicing planner.

Electric utility analysts and interest group advocates undertook a parallel search for an alternative to rational planning throughout the 1980s, though that search was scarcely connected to the one undertaken by urban planners. Though disagreeing on particular solutions, virtually all analysts and advocates agreed that technological and economic changes had undermined the rationale for having a single "natural monopoly" supply electric power to an entire region. The industry was, they agreed, ripe for fundamental restructuring. But restructured to what? And guided by whom or what? Looking ahead and conceiving of electricity merely as a commodity, many analysts argued that utilities should be allowed to compete with one another or with other large new entrants into the power generation business. Accordingly, they debated the merits of total deregulation of power generation, deregulation of new generating plants only, deregulation of wholesale power exchanges between utilities, "incremental asset deregulation," and federal preemption of utility regulation. Still other analysts and advocates tended to define electricity more broadly, either as a community-based public service or as a product of small and decentralized power systems linked into regional networks. Accordingly, they tended to debate the merits of regionalized power planning and regulation, "competitive bidding," "least-cost utility planning," and municipalization (Fenn 1984; Rudolph and Ridley 1986; Hyman 1988).

Thus by the mid–1980s, planners who sought to guide the development of cities, of public policies, and even of regional electric power systems seemed to be plummeting quickly down from the central plateau of modernist planning. Modernist planners had believed that rational design based on subject-centered reason would produce rational cities and rational policies. Ironically, however, that effort to impose Reason and Order on cities and policies helped to fragment cities both socially and economically. Planning itself seemed to have become chaotic and fragmented (Dear 1986), leaving planning theorists deeply divided over what planning was and how it could best be performed (Beauregard 1990; Klosterman 1992). Its primary "text" (the modern industrial city) seemed to be fragmenting and dispersing, and its many "readers" offered conflicting and often contradictory accounts about what planning and its object of attention should be. Seeking to create a future in which man would be "in total control," modernist planners had helped to create a world in which humanity teetered at the brink of nuclear annihilation and environmental destruction. Claiming to guide change, urban and regional planners found themselves trying to chase, correct, and modify changes that had been produced by the market economy, racial biases, and gender-based presumptions. Contrary to the presumptions of modernist planning, planning turned out to be both technical and political, with its policies and designs being shaped in part by conscious intent but also in part by the politics, economics, and race and gender relations of twentieth-century America. Like other disciplines and professions, planning had entered a postmodernist, "multicultural" era (see Berman 1992; Hunter 1991; Rosenau 1992). What could planning be in such a context?

Interlude

A Scholar's Office in Iowa City

I walk into the room, flick on the light, and sit down at my metallic grey desk. "Another cold and dark winter day," I think. "But at least this gives me a chance to concentrate on my research project." I stare at the tons of research materials that border and define my office. Shelves full of books about planning theory, planning history, and electric power. Journals everywhere: the Journal of the American Planning Association, Policy Sciences, *the* Journal of Policy Analysis and Management, Science, Public Utilities Fortnightly, Power Line, *and many others. Dozens of governmental reports. Files full of articles from other journals. Pacing nervously, I wonder how I will be able to construct a research design that meets the best standards of academic rigor. Somehow, I do not feel comfortable at the thought of simply analyzing given data or constructing and testing models as if the world speaks unambiguously through them. Staring out the window, watching the snow fall gently to the ground, I find myself remembering my experiences as a planner and the untold thousands of conversations and readings that helped comprise those experiences. "No," I think, "data and models are important parts of the story. But they're not the whole story." Language, rhetoric, and discourse, I had become convinced, form the heart of planning practice. Somehow I have to convey that to my readers.*

2

The Argumentative or Rhetorical Turn in Planning

In the late 1800s, Daniel Burnham encouraged his followers to "make no little plans," to cure the ills of the modern industrial city with the timeless and universal principles of truth and beauty. Ninety-five or so years later the city seemed to be fragmenting and dispersing, both as a physical entity and as the primary object of the planners' attention. As its primary object of concern seemed to disappear, "planning" seemed to fragment and disperse as well.

What became of the city? Radically transformed by public policies and private investment decisions, the city's boundaries have become so blurred as to make it difficult to speak coherently of the city having a center, an inside and an outside.[13] The city has had to reach farther and farther afield to obtain the resources required for its sustenance and growth and to dispose of its products and wastes. It has become less like an isolated work of art, or machine, or organism, and more like a node in a global-scale network (or web) of links and flows (Noyelle and Stanback 1983; Castells 1991; Abu-Lughod 1991). Los Angeles, for example, can be understood to include the flow of oil from the Persian Gulf, the flood of water from the Sierra Nevada Mountains, the smog pouring into the deserts of Arizona, and high-level nuclear power wastes destined for long-term geological disposal in Nevada.

And what of the planner and planning? Once a physical design-based activity focused on the city, the work of planners ("planning") gradually fragmented and dispersed to include people of many disciplines who were interested in state and federal and corporate policy making. Focusing more on policies, these planners became enamored with the rational model. But even that model deconstructed in the 1950s and 1960s. In the sixties alone, many planners arrived at one or more of the following conclusions: (1) administrators did not decide in accord with the rational model; (2) planners should advocate the interests of low-income and other groups; (3) the political economy itself is corrupt, and planners should help to emancipate people (African Americans, women, Native Americans, gays) and nature from oppression.

Now, in the mid-1990s, we planners find ourselves adrift in a sea of shifting images of who we are and what our proper focus should be. Modernist planning has become just one of the many rhetorics that planners speak, and "the planner" has become a decentered network of diverse rhetorics. Design-oriented planners imagine orderly and beautiful spaces, comprehensive planners devise schemes for using land efficiently, rational planners seek to discover the best way to solve predefined problems or achieve predetermined ends, advocacy planners try to represent disadvantaged interests, policy analysts seek to provide expert advice to powerful decision makers, feminists seek to focus attention on gender issues, African Americans decry the pernicious effects of racial bias, environmentalists argue that there are natural limits to population and economic growth, and so on. Having fallen from the central plateau, we now find ourselves teetering at the brink of a multicultural "postmodern abyss" (Beauregard 1991).

Losing our balance, fearful of the fall, scarcely able to talk with one another, we wonder what to do.

It would help, I think, to begin by rejecting the assumption that there is such a thing as "the city." Rather, we can think of the city (that node in a global-scale web of links and flows) as being like a text that can be read and interpreted in diverse ways. As the urban sociologist Janet Abu-Lughod puts it, "people who live in the same city may, in fact, be inhabiting very different worlds, not only because cities contain a variety of physical and social settings within their boundaries but also because people perceive and interact within these settings in significantly different ways—they 'inhabit' different cities" (1991, 323). Furthermore, the once powerful distinction between "urban" and "rural" has lost much of its meaning. "Urban" areas have dispersed to become complex metropolitan regions that blend city centers, suburbs, outlying exurban zones, and special-function towns (Abu-Lughod 1991). And rather than being spatially determined, as planners once thought, communities can now better be understood as social constructions (Fischer 1982).

It would help as well to abandon the idea that "planning" and "the planner" can be defined authoritatively and unambiguously. Rather we can think of a "metatext" of planning in which planners and others (those who consciously try to influence activities explicitly labeled as planning) write second-order texts (e.g., plans, analyses, forecasts) that interpret and construct the city and planning. Others then read and interpret those second-order texts. Already actively constructing what "the city" means to them, those readers actively construct diverse stories about its development path by drawing on fragments already provided by planners and others. They create multiple reading paths through the "metatext" of links, nodes, and flows, and they argue over what the city-text is and should be, and—to the extent that they are able—proceed to reshape that city-text in accord with their readings.

Thus planning can be understood as a fragmented and heterogeneous mix of stories and storytellers in which no one rhetoric has a *prima facie* right to be privileged over others. Rather, planning stories and storytellers exist and relate to one another in an interconnected web, a web of partial truths. In this web of partial truths, it becomes important to ask what makes one interpretation more persuasive than any other. It becomes important to attend to the rhetorical dimensions of planning.

The Rhetorics of Planning

What is rhetoric? According to the conventional view, it is the use of style to manipulate or seduce others into behaving or thinking in some desired way (Gronbeck 1983; McGee and Lyne 1987). In this sense, rhetoric is ornament, display, or "mere rhetoric" that threatens to reduce all judgments to immediate persuasive effect rather than sound intellectual argument. That's just rhetoric, one might say; all style and no substance.

This traditional view has recently been challenged, however. Throughout the sciences one can detect a broad turn away from positivism, modernism, and objectivity and towards a concern for the ways in which language, discourse, and rhetoric construct society and our knowledge of it (see McCloskey 1985; Nelson and Megill 1986; Nelson, Megill, and McCloskey 1987; Bazerman 1988; Klamer, McCloskey, and Solow 1988; Majone 1989; Simons 1989; Hunter 1990; Wetlaufer 1990; Simons 1990; Throgmorton 1991; Landow 1992; Rosenau 1992; McCloskey 1994). This turn toward rhetoric (in its deeper sense) is connected with a much larger scholarly conversation about poststructuralism, postpositivism, critical pluralism, hermeneutics, and critical theory.[14]

What then is rhetoric if not mere ornament? According to Nelson,

Megill, and McCloskey (1987), it is persuasive discourse within a community, or honest argument directed at an audience; it is "the quality of speaking and writing, the interplay of media and messages, the judgment of evidence and arguments" (ix). Accordingly, one who investigates rhetoric within a discipline would analyze actual arguments among scholars. Students of rhetoric would begin with texts and take into account the roles of various audiences for inquiry; they would explore how backings are shared, the extent to which warrants are accepted, and why. They would be concerned with figures of speech, arguments, and other devices of language authors use to persuade audiences in particular disciplines, and they would assume that words ("planning" and "the city," for example) have no meaning apart from their use in particular communities.

Leith and Myerson (1989) complement this view when they argue that a minimally satisfactory rhetorical approach would build on three principles:

- Arguments and stories are always addressed to someone else. To persuade, those stories and arguments have to take that "someone else" into account. Therefore, audience is an important concept.
- All utterances can be seen as replies to other utterances. Thus "to argue is not merely to put forward a view, but also to speak, or write, *in the awareness of a differing or opposing view*" (85).
- The meaning of the utterance will always go beyond the conscious control of the author. So we might think about the "play" of meaning, and about how audiences (or readers) construct the meaning of utterances.

A rhetorical approach to planning would, therefore, emphasize the importance of trying to persuade specific audiences in specific contexts to accept proposed explanations, embrace inspiring visions, undertake recommended actions, and so on. But it would also acknowledge that *such persuasive efforts take place in the context of a flow of utterances, replies, and counterreplies.*[15]

These new rhetoricians tend to focus on conversations within disciplines. Though fruitful, such a focus is too narrow for a field such as planning in which authors/speakers can occupy numerous roles and direct themselves to numerous audiences. So, to understand the rhetorics of planning we first need to have a sense of who the audiences for planning are. At the risk of oversimplifying a very complex situation, I want to propose a simple conceptual framework, a framework that

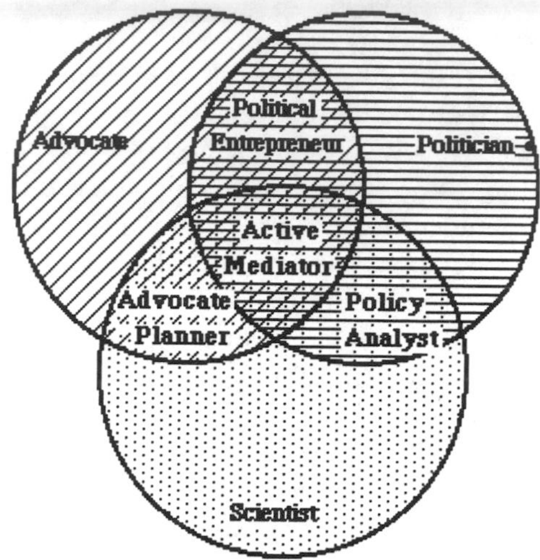

FIG. 2.1 **Conceptual Model of the Rhetorics of Planning**

initially accepts but then later disrupts several core assumptions of modernist planning.

Let us initially assume that individual planners can occupy one of three roles: scientist, politician, or advocate (see fig. 2.1).[16] Further assume that these planners have three parallel audiences for their work: scientists, public officials, and lay advocates and that the planners' task is to persuade each applicable audience to cite, use, or trust their work. That is not an easy task, for those audiences vary dramatically.

Scientific planners seek to increase the cumulative stock of knowledge in a particular discipline and define planning as primarily a scientific, technical activity. Guided by a theoretical framework provided by their home discipline, they seek to describe, explain, and predict phenomena in a theoretically coherent way and then to use the resulting law-like generalizations and associated predictions to produce desired "outcomes" (MacIntyre 1981). From their point of view, arguments over values cannot be decided rationally. The best such planners can do is identify "the best" or most "rational" way of attaining goals that have been chosen through "irrational" processes.[17]

Scientific planners tend to address technical peers (primarily through technical journals and reports), and their rhetoric is dispassionate. As Ziman puts it, scientists write a paper

as if it were addressed to a hypothetical, very skeptical reader, who is already very well informed on the subject, and might therefore form the spearhead of

> critical opposition. It is couched in formal technical language, thus indicating the professional competence of the author. Other research bearing on the subject is religiously cited, both to authenticate the basic premises of the investigation and to indicate that the author is thoroughly familiar with all the backgound material. The theoretical arguments and experimental results are expressed impersonally, in the passive voice, as if to emphasize the objectivity and disinterestedness with which the research was undertaken, and the conclusions are given quasi-logical weight, suggesting the rational necessity of this particular outcome. (1984, 62)

Similarly distinguishing between facts and values, other planners (*advocates*) define planning in the moral language of rights and wrongs. Whether they are promoting the rights of women, gays, African Americans, consumers, the unborn, future generations, or the natural environment, advocates speak of rights, justice, and (in many cases) freedom from oppression. It is in this spirit that Michael Brooks (1988) appeals for "a return to the utopian, visionary, reformist spirit that previously fueled the [planning] profession" (241) and claims that we bear "critical responsibilities . . . for the well-being of all who reside in the communities we purport to serve" (246). Through words and deeds they try to transform perceptions, express demands, and mobilize action. They tend to write and speak in the vernacular, and they communicate with the public through community meetings, slogans, songs, pamphlets, paperbacks sold in bookstores, and routine face-to-face interaction. When necessary, they use emotional appeals and symbolic actions (e.g., demonstrations and marches) to create a groundswell of public support for favored policies and programs (Stewart, Smith, and Denton 1984; Hunter 1991). They believe that important public decisions should not be made without their informed participation and that pluralist politics helps maintain a system of oppression. They tend to reject compromise as capitulation.[18]

Perceiving this radical moral dissensus, a third group of planners (*politicians*) define planning as just another part of the "game" of politics. According to this view, important decisions are made not by planners and their "rationality" or "utopian, reformist spirit" but by elected officials who want to "get things done" (produce desired consequences) in a context of incremental change, fragmented authority, legal constraints, and self-interested competition for electoral position, power, and survival (Matthews 1988).[19] Seeking to gain and hold political power, they try to show that they are qualified for elected office, are like their constituents, and empathize with their constituents' problems (Fenno 1978). They speak of friends, enemies, alliances, favors, deals, loyalty, reputation, spin (managing expectations and reactions to performance), and positioning (presenting a desired self-image to others). In this "political" view the past is far from irrelevant—it is the rumpled slate on which planners have to draw—and if planners want to "master change" (McClendon and Quay 1988) they have to find a way to be "close to power" (Lucy 1988). They have to help elected officials emphasize good trends and ignore bad ones, claim credit for what is good and shift blame for what is bad, and criticize others for past decisions and for unpopular trends and events, regardless of the extent to which the others actually caused them.

As this brief review indicates, the contrast between the rhetorics of "explaining phenomena in a theoretically coherent way," "putting the right spin on things," and "demanding rights" is stark.[20] I want to suggest that the central problem with this trimodal rhetoric of modernist planning is its assumption that reasoning about truth, morality, and effective action can be abstracted from its context; that is, that any person will, if reasoning properly, arrive at the correct understanding of what is true or right or feasible. This assumption ignores the importance of audiences and speaking in the context of a flow of utterances, replies, and counterreplies. As practicing planners rediscover time and again, "truth," "goodness," and "effectiveness" are contestable claims that must be justified and defended before diverse audiences and in terms of what has been said before. Thus modernist planners have to persuade diverse audiences to accept and act upon the planners' claims. At this point planning takes a late-modern turn: taking up the rhetorical challenge of persuading diverse audiences, late-modernist planners have tried to articulate new roles for planners, roles that explicitly link passion, reason, and power.

The conversations within these disparate audiences can be thought of as "normal discourses" (Rorty 1979) within "interpretive communities" (Fish 1979). Each of these communities has "an agreed-upon set of conventions about what counts as a relevant contribution, what counts as answering a question, what counts as having a good argument for that answer or a good criticism of it" (Rorty 1979, 320). When disagreements arise within these communities, members try to persuade one another with rhetoric authorized by their "agreed-upon conventions." However, when communication between interpretive communities is required—that is, "when someone joins in the discourse who is ignorant of these [normal] conventions or who sets them aside" (320)—members of these interacting communities have to engage in "abnormal discourse." At this point the criteria for persuasion become much more ambiguous. Hermeneutics becomes necessary.

Hermeneutics can be thought of as the study of an abnormal discourse from the point of view of some normal discourse (Rorty 1979). As MacIntyre (1988, 350) puts it, "There is no standing ground, no place for enquiry, no way to engage in the practices of advancing, evaluating, accepting, and rejecting reasoned argument apart from that which is provided by some particular tradition or other." Thus scientists who seek to understand the political community would approach that community from the traditional perspective of the scientist. They would provisionally concede authority to the political, seeking to understand it better. At this point, dialogue between the two communities would become crucial; scientists would have to be prepared to accept their own fallibility and to discover the real strength of the political perspective. Successful dialogue would produce a shared understanding, or "fusion of horizons" (Bernstein 1983; Geertz 1983; Gadamer 1991), that reflects a transformation of the initial positions of both communities. They would construct a new common language.

The experiences of engaging in "abnormal discourse" and trying to "fuse horizons" are common ones for planners. As they become more experienced, move from one role setting to another, or delve more deeply into planning problems, planners must engage a wider range of planning audiences. In doing so, they tend to act hermeneutically and thereby become more like the audiences they try to persuade. Over time, new forms of normal discourse develop between some members of each of these communities. The three initially "pure" impulses tend to shade into one another and thereby create three new, and more realistic, roles for planners.

As some scientists, politicians, and advocates interact, the incoherence of "abnormal discourse" becomes less confusing. Interacting members gradually "fuse horizons" and construct new interpretive communities: policy analysts, advocacy planners, and political entrepreneurs. To be persuasive within these new interpretive communities, planners have to use rhetorics appropriate for (and expected by) each of those communities. Each of these new rhetorics, however, contains conceptual ambiguities and ethical tensions.

One role might be termed client advocate (Jenkins-Smith 1982), strategic planner (McClendon and Quay 1988), or *policy analyst* (Wildavsky 1987). Blending science and politics, these planners try to persuade elected and appointed officials to use the planners' advice (see Majone 1989). How they proceed depends in part on whether they initially see themselves as political or scientific planners. Scientifically oriented analysts (e.g., Stokey and Zeckhauser 1978) emphasize the importance of methodological rigor, but they also stress the importance of brevity, clarity, and timeliness. Politically oriented analysts (e.g., Wildavsky 1987; Meltsner 1985) stress the importance of addressing problems that the client has the authority to solve and using language and concepts the client can understand, or—as Wildavsky (1987, 16) puts it—constructing "problems that decision-makers are able to handle with the variables under their control and in the time available." From this point of view, analysts must be willing to delete material from reports and do whatever else—within vague limits—is necessary to help gain acceptance of their advice and their clients' policies (Meltsner 1985).

A second new role might be termed partisan analyst (Lindblom 1980), *advocate planner* (Davidoff 1965), or issue advocate (Jenkins-Smith 1982). Disclaiming value-neutrality, these planners use rigorous scientific techniques in support of a chosen set of values or interests. As Davidoff (1965, 331) puts it, "the planner should do more than explicate the values underlying his prescriptions for courses of action; he should affirm them; he should be an advocate for what he deems proper." According to this view, advocate planners help clarify and express their clients' ideas, help inform the public about the range of available alternatives, point out biases in other plans, and ground the evaluation of plans in a clearly articulated (and passionately held) set of values.

The third new role might be termed mobilizer (Rabinovitz 1969) or *political entrepreneur* (Mollenkopf 1983). Blending advocacy and politics, and relying heavily on "ordinary knowledge" (Lindblom and Cohen 1979), these nontechnical planners bring diverse constituencies (or advocacy groups) into new coalitions, which in turn push for the expansion of programs and provide organizational and political support for the political entrepreneurs who established them. According to Mollenkopf (6) the "'political entrepreneur' . . . does not simply play by the rules of the game, but attempts to win the game by changing them. Using government to create new beneficiary groups, political entrepreneurs create supportive new constituencies." Robert Moses might be the best example of such a planner (Caro 1974).

Emergence of the policy analyst, advocate planner, and political entrepreneur roles demonstrates that rhetoric can help bind interpretive communities together. But each of these rhetorics incorporates conceptual ambiguities and moral tensions. By acting as policy analysts, planners can increase their ability to influence decision making. But they do so at the risk of losing legitimacy in the eyes of particular rights-oriented communities (Levine 1982). Scientists will accuse analysts of being "too applied" or contaminated by politics, and lay advocates will accuse analysts of being untrustworthy mercenaries who systematically perpetuate wrongs; e.g., racial segregation (Feld 1989). Occupying the role of advocate planner earns the planner greater credibility and legitimacy with particular communities, but it does so at the risk of losing credibility in the eyes of elected officials. Advocacy planners are also open to the scientist's charge that values contaminate their work and to the analyst's charge that they are substituting their own political values for those of the broader public (as represented by elected officials) (Hill 1989). Planners who assume the role of political entrepreneur are more likely to get things done in a way that satisfies both elected officials and particular communities, but they do so at the risk of being accused by scientists of producing decisions that are technically incompetent, ineffective, and guided by emotion.

It is in the mixing of roles and audiences, in the effort to make practical sense out of being caught between three audiences and three rhetorics, in the effort to "fuse horizons" and thereby "do good, be right, or get things done," that the complex rhetorics of planners emerge. This effort has led to the development of the three late-modern rhetorics of policy analysis, advocacy planning, and policy entrepreneurship. But, as we have seen, each of these blended rhetorics expresses difficult conceptual ambiguities and moral tensions. Can those tensions and ambiguities be eliminated in practice?

One possible solution is to combine those diverse rhetorics into some kind of grand synthesis that is "above the fray." An appealing dream, but a dream nonetheless. Indeed, the very idea of synthesizing such diverse rhetorics demonstrates the power of modernist planning. Such a communicative synthesis is simply another effort to rise to the heights of the central plateau from which we planners have fallen.

A second alternative is to take Leith and Myerson's point about prior utterances seriously. Scientists might speak, but it is politicians and advocates (as well as other scientists) who listen and respond, putting their own twists on the scientists' words. As politicians reply, counterreplies come from advocates and scientists (as well as other politicians). Though modernist planners would prefer otherwise, this messy and often chaotic interplay of diverse rhetorics cannot be reduced to a procedural flow chart which someone calls "the planning process." Rather, the interplay can more meaningfully be characterized as a flow of utterances, replies, and counterreplies that seek to persuade diverse audiences. This flow, together with the setting in which it occurs and the real flesh and blood characters who seek to direct the flow through oral and written texts, might best be thought of as a story. So it is to stories and to storytelling that we should turn if we want to conceive of a form of planning that is appropriate for the fragmented and multicultural world in which we live and work.

Planning as Persuasive and Constitutive Storytelling about the Future

To draw attention to the importance of storytelling in planning is not in itself novel. Over ten years ago, Martin Krieger (1981, x) wrote that "plans are works of art and artifice and experimentation" and that "they share the literary and analytic virtues." Advice is expressed in "reasonable and justifiable" stories, he claimed, and planners can become more effective storytellers by mastering appropriate narrative forms, suitable character development, and effective stylistic devices. In the years since, several other scholars have joined Krieger in discussing the importance of storytelling in planning, analysis, and the social sciences; e.g., Payne (1984), McGee and Nelson (1985), Neustadt and May (1986), Kaplan (1986), Roe (1989), Marris (1990), McCloskey (1990), Beauregard (1991), Mandelbaum (1991), Hoch (1992), Cronon (1992), Forester (1993), and Ferraro (1994).

I want to build on this work. In the remainder of this chapter, I will explain why storytelling deserves attention as an important part of planning. This explanation will draw heavily on arguments presented by Alasdair MacIntyre (1981), Martha Nussbaum (1990), and Walter Fisher (1989). Having suggested good reasons for attending to storytelling, I will then discuss what makes one planning story more persuasive than another. This will lead me to suggest that planning can be likened to good fiction and that planners are future-oriented storytellers who write persuasive and constitutive texts that other people read (construct and interpret) in diverse and often conflicting ways.

Most readers of this book probably enjoy reading good stories. But that does not mean that storytelling should be an important part of planning scholarship or practice. Why should it be? MacIntyre starts to answer this question by claiming that man is "essentially a storytelling animal . . . a teller of stories that aspire to truth" (216). In his view, we humans engage in "enacted narratives" that have beginnings, middles, and ends, that embody reversals and recognitions, that move toward and away from climaxes, and that might contain digressions and subplots. What is more, to understand what other people are doing, we have to place their action into a context of narrative history:

"the act of utterance becomes intelligible," he writes, "by finding its place in a narrative" (210). But what of the future, of how our stories will evolve over time? Here MacIntyre stresses three factors. First, no individual is able to act unilaterally. Rather, what any person is able to do and say intelligibly is "deeply affected by the fact that we are never more (and sometimes less) than the coauthors of our own narratives" (214). The second and third factors flow from that coauthorship; the future is both unpredictable and partly the result of the intentionality of human action. "Like characters in fictional narratives we do not know what will happen next," he writes, "but nonetheless our lives have a certain form which projects itself towards our futures" (216).

Drawing on MacIntyre, then, we can think of planning as an enacted and future-oriented narrative in which the participants are both characters and joint authors. Though these authors may construct their own stories, they do so as characters in other stories, and they take part in actions that are not entirely of their own making.

How can the nature of such an extended, future-oriented, enacted narrative best be conveyed? One possibility that many of the readers of this book probably favor is to rely on a plain, clear, general, nonnarrative style. Style, in this view, is largely decorative. Good scholars should skip the decoration and stick to cold, hard argument; that is, to detached, purely intellectual reasoning. Nussbaum (1990) firmly disagrees with this point of view, arguing that style is not separable from content. She argues to the contrary that "the telling itself . . . expresses . . . a sense of what matters and what does not" (5) and that many views of the world cannot be fully and adequately stated in the language of conventional nonnarrative prose. Rather, some views of the world can only be fully and adequately stated in ways that are more complex, more allusive, more attentive to particulars; in a word, through stories.

At this point, Nussbaum's argument becomes rather complex. Novels (or stories), she argues, convey truth about human lives in four ways that nonnarrative analyses do not. First, they draw attention to the *noncommensurability of the valuable things.* Not all valuable things can be reduced to a single metric, as in benefit-cost analysis for example; rather, characters often have to choose between two qualitatively different actions or commitments. Second, novels emphasize the *priority of perceptions;* that is, the ability of characters to take into account, as general fixed principles do not, three salient features of their particular situations: (1) new and unanticipated features (the unpredictability that MacIntyre refers to), (2) the context-embeddedness of relevant features (the relationship between those features and the "enacted narratives"), and (3) the ethical relevance of particular persons and relations. Third, novels stress the *ethical value of the emotions.* Here Nussbaum challenges the idea that emotions are unreliable, animal, seductive, and lead us away from the "cool reflection that alone is capable of delivering a considered judgment." She suggests instead that emotions (we planners might say "politics") have a cognitive dimension; they are "discriminating responses closely connected with beliefs about how things are and what is important" (40–41). Last, novels highlight the *ethical relevance of uncontrolled happenings.* This implies that good planning is not simply a matter of good technique, but that the quality of planning is intimately related to the planners' reponses to actions and events that occur during plan "implementation." "By showing the mystery and indeterminacy of 'our actual adventure,'" she writes, "they [novels] characterize life more richly and truly" (47) than does conventional nonnarrative prose.

So, we can think of planning as an enacted, future-oriented, narrative in which the participants are the actors and joint authors, and we can think of storytelling as being an appropriate style for conveying the truths of planning action. What, however, should be done when planning stories overlap and conflict? How can planners and others (who are simultaneously coauthors of and characters in each other's stories) decide which planning story is more worthy of the telling?

Fisher (1989) begins to answer this question by introducing the concept of *narrative rationality.* According to this view, humans are "storytellers who have a natural capacity to recognize the coherence and fidelity of stories they tell and experience" (24). We test those stories in terms of the extent to which they "hang together" (*coherence*) and in terms of their truthfulness and reliability (*fidelity*). We test the coherence of a story in three ways: in terms of its internal, structural coherence; in terms of how well it deals with issues and counterarguments appearing in competing stories; and in terms of the reliability of its narrators and characters. The only way to bridge the gap between rival stories, he suggests, is to tell stories that "do not negate the self-conceptions that people hold of themselves" (75). The fidelity of a story can be assessed by applying "the logic of good reasons," by which Fisher means elements that provide warrants for accepting or adhering to the advice addressed to an audience (48). Thus fidelity concerns the "truth quality" of a story, whereas coherence concerns the sequence of thought and/or action that formally constitute the story, and technical logic becomes meaningful and useful only in the context of a story as a whole. Kaplan (1986) produces criteria similar to those of Fisher when he suggests that good policy analytical stories have to meet five criteria. They have to be true and rich (in the sense of encompassing all the realistically possible true stories and showing why other stories are either false or inapplicable), and like other narratives, they should be internally consistent, congruent, and unified.

Following Fisher, then, we might conclude that planners and others can compare planning stories in terms of their narrative rationality. Such a conclusion remains unsatisfactory, however. One can easily imagine a situation in which two planning stories, both of which are coherent and truthful, compete for attention. What, then, makes one coherent and truthful narrative more worthy than another?

I want to suggest that the answer to this question lies, in part, in the persuasiveness with which we tell our stories. More specifically, I want to suggest that if planning is a form of persuasive storytelling, then planning can be likened to good fiction.[21] Planners can be regarded as *authors* who write *texts* (plans, analyses, articles) that reflect awareness of differing or opposing views and are normally *read and interpreted* in diverse and often antagonistic ways. But planners do not simply write texts; they are also characters whose forecasts, surveys, models, maps, and so on, act as *tropes* (figures of speech and argument) in their own and in other people's persuasive stories.

Think first of planners as authors. As authors, planners have to *emplot* (or arrange and shape) *the flow of action.* And if they do it well, they will arrange and shape the flow of action "as performed by characters, realized in their details, seen in their atmosphere, from a unique point of view, through the imagery and rhythm of the language" (Burroway 1987, 342–343).

Continuing to draw on Burroway, I want to suggest that persuasive planners follow a few basic principles when writing such stories. Perhaps most important, good authors build *conflict, crisis, and resolution* into their narratives, such that key protagonists are somehow changed or moved significantly. Doing this means incorporating what otherwise would be conflicting policy stories into the authors' own and having one of them (or some new story) win out in the end. Lacking such resolved conflict, readers will wonder why they should bother reading the story. Second, good authors build *characters* into the narrative, characters who are interesting and believable and whom readers (many of whom are also characters in the stories) care about. That means writing about interest groups, public agencies, ordinary people, elected officials, and others who play important roles in decision making on particular issues. Otherwise, readers will be less inclined to care how the story turns out. Third, good authors place the action in its rightful context. That means acknowledging the *settings* in which those characters come into conflict. Regulatory commission hearing chambers, for example.

Fourth, good authors adopt an appropriate *point of view.* To do so they have to ask, both for themselves and their characters, Who is standing where to watch the scene? Who is speaking? To whom? In what form? At what distance from the action? With what limitations? Once having adopted a point of view, good authors normally stick to it.[22] Last, good authors use the *imagery and rhythm of the language* to express a preferred attitude toward the situation and its characters.

Embedded in the imagery and rhythm of language, at the heart of persuasion, are *tropes;* that is, any literary or rhetorical device, such as metaphor, metonymy, synecdoche, and irony, that involves using words in other than their literal sense (Quinn 1982). It implies a turn toward something, a turn induced by the device itself, and can be either a figure of speech or a figure of argument. For example, when we talk about "the fifth floor of City Hall" but mean the mayor, we are using *metonomy* (substituting the name of one object for that of another to which it is related or of which it is a part). When we use surveyed samples to represent entire populations, we are using *synecdoche* (substituting a part for a whole). When we use computer models to simulate electric power usage patterns, we are using *metaphor* (an implied comparison between two things of an unlike nature). When we confidently forecast future demand for electric power while knowing that our prior forecasts proved far off the mark, we are being—whether we know it or not—*ironic* (conveying a meaning opposite to that which is literally intended). When we weave these tropes together into a plan or analysis, we are engaging in persuasive discourse.

So planners are authors who write persuasive stories about the future. As they write their stories, attempting to follow principles such as those outlined by Burroway, they must be sensitively aware that many of the characters in their stories are also *readers and critics* of those stories (Mandelbaum 1990). This suggests that persuasive planners also need to be broadly familiar with the relationship between readers and texts.

According to the "new criticism" which ruled the literary world from the twenties till the sixties, the meaning of a text (a plan or the present chapter, for example) lies neither in the author's intentions nor in the readers' response. Rather it lies objectively in the written text itself, and readers need to be taught how to read correctly; that is, through a "close reading" that is "objective" and "scientific" (Freund 1987). Few literary critics adhere to that view anymore. For the past twenty years or more, they have been shifting their attention to *reader-response* (see Fish 1979; Freund 1987; Tompkins 1980). According to reader-response theory, the distinction between the author's "text" and the readers' construction cannot be maintained, or as Tompkins (1980) writes, "reading and writing . . . finally become distinguishable only as two names for the same activity" (ix). Readers assign different meanings to key terms, fill gaps in the original "text," and choose to read either with or against the "text." Fish (1979) concludes that it is "interpretive communities, rather than either the text or the reader, that produce meanings" (14) and that the business of criticism is "not to decide between interpretations by subjecting them to the test of disinterested evidence but to establish by political and persuasive means . . . the set of interpretive assumptions from the vantage of which the evidence (and the facts and intentions and everything else) will hereafter be specifiable" (16). Lacking a privileged point of view, all authors and critics must persuade in the face of anticipated objections.

Note two critically important planning implications of this notion that readers help to construct texts. Even the most abstract, theoretical, and scientific texts will—to the extent that someone tries to apply them to the world of practical action—be read as possessing political and normative content. To the extent that such content is not overtly present, savvy readers will fill in the blanks themselves (see also Yannow 1993). They will read the "text" as part of a larger story, and they will assume that the original author left that larger story out for reasons consistent with the readers' own version. Note also that planners and the characters (potential readers) in their stories form an interpretive community. Though they might disagree with one another's stories, they understand them. Readers outside the community might not even recognize the meanings attributed to specific words or to stories as a whole.

Planners are, according to this view, authors of persuasive and future-oriented texts (plans, analyses, articles) that reflect differing or opposing views and that can be read and interpreted in diverse and often conflicting ways, and planners achieve persuasiveness by attending to key principles of fiction writing and reader-response theory.

Such future-oriented storytelling is not, however, simply persuasive. It is also *constitutive:* to use James Boyd White's (1984, 1985) terms, the ways in which planners—and others involved in the process of planning—write and talk shape community, character, and culture.

Consider community first. How "we" (as authors) write and talk shapes who "we" (as a temporary community of authors and readers) are and can become. As authors, planners create "ideal readers" whom actual readers may or may not be or choose to become. Thus when they write and speak in the passive voice, use technical jargon and acronyms, and express themselves in convoluted syntax, planners create ideal readers who assume that behavior is determined by Godzilla-like impersonal forces that only a few people are able to understand and fewer still can control (Hoch 1992). The planners' rhetoric can help to reproduce existing communities or create new ones (see Gottlieb 1993).

Consider character as well. By their choice of tropes, planners create images of the kinds of characters they are or want to become (see Frug 1988; Schram 1993). Take the metaphor of the city as machine. To speak of the city as if it were a machine leads planners to think of themselves as akin to the scientists, engineers, and technicians who make machines run more efficiently. And it leads them to think of their readers as passive parts of the machine. Or consider econometric models of electric power demand. Such models are metaphorical ways of talking about who uses electric power, how much power, and why. As metaphors, such models induce electric power planners to think of themselves as experts who can predict electric power consumption behavior. Simultaneously such metaphors lead planners to think of political officials and interest groups as irrelevant agents who (since they do not appear in the models) cannot (and should not) influence those behaviors.

Recognition that future-oriented storytelling can be both persuasive and constitutive should lead planners to ask a fundamental ethical question: what kinds of communities, characters, and cultures do they want to help create? Asked differently, what ethical principles should guide and constrain the planners' efforts to persuade their diverse audiences? If planning involves a fragmented and heterogeneous mix of persuasive stories and storytellers, any reasonable answer to this question must presume a diverse array of rhetorics and ethics. So let me tentatively suggest five guiding principles that take that diverse array into account.

The first principle is that planners (and others engaged in planning) should think of themselves as characters in a larger story that they are helping to construct, and that they should strive to act in a manner that is consistent with the characters invoked by their story. Planning is persuasive storytelling about the future, and competing stories abound. Each of those competing stories constructs characterizations that are consistent with its overall plot. Given this narratival pluralism, planners must learn to act and speak as characters who appear in their own stories. To do otherwise, to violate the *ethos*

I currently serve as an elected member of the Iowa City City Council. As a city councilor, I am acutely aware that many people characterize me and other elected officials as "craven politicians" who are beholden to "special interests." I could choose to act in a manner that is consistent with that characterization, but I

choose not to, preferring instead to act as a thoughtful elected official who seeks to promote a progressive vision of the future while also being willing to negotiate and compromise when necessary.

of one's own character by acting as some other story would require, is to embrace and validate the other story rather than one's own. If the story one embraces says that planning is primarily a scientific enterprise, then one should act like a scientific planner.

The second principle is to recognize that planners (even the most scientific ones) are also the authors (or at least coauthors) of at least one of those stories and hence that as authors they must adopt a point of view. That point of view initially derives from the normal discourse of the interpretive community of which they are a part.

Here I want to be clear about the point of view I adopt in this book. I am a scholar, and as such I must strive to attain and maintain credibility with other scholars. That means accurately citing sources of quotes and data, cross-checking claims to verify points of agreement about alleged facts, connecting my arguments to related research, and exposing my arguments to scholarly peers for review and critique. However, I am also an advocate. Like many other planners (e.g., Davidoff [1965]; Friedmann [1987]; Forester [1989]), I would argue that capitalist democracies bias institutional processes (the rules of the game) in favor of wealthy businesses and individuals. Moreover, in the particular context of electric power planning, I am persuaded that those processes have been biased against using energy more efficiently and in a more ecologically sustainable way. I am, therefore, predisposed to advocate the values, interests, and voices of less-advantaged people in our society. However, my advocacy for particular voices does not authorize me to violate my scholarly commitments or to ignore inherent weaknesses in the moral or technical underpinnings of arguments presented by consumer and community groups. Some readers are likely to fault me for adopting the point of view just presented, and I accept their critical reading as part of the abnormal discourse of contemporary planning and policy research.

As authors with membership in an interpretive community, planners actively construct the flow of their future-oriented narratives. Thus the third principle is that planners have intentions, and that they should seek to accomplish those intentions in part by how they emplot the future course of their narratives. That means shaping the readers' attention and expectations rhetorically (Forester 1989). This active construction of narratives is unavoidable. Even scientific planners do it. Later in the book I refer to this active construction as imagining (or inventing) and articulating a path into the future and then attempting to persuade diverse audiences that the path is both desirable and feasible.

The fourth principle is to recognize that planners are at best coauthors of their own stories while also being characters in someone else's story. So planners should warmly embrace the need to engage their audiences (the other coauthors) in dialogue.[23] Regardless of the position from which planners start, they must inevitably encounter diverse audiences, audiences that often hold positions that differ quite radically from the planners' own. Planners should listen to those contending stories, learn what roles they assign, and relate those stories and characterizations to the planners' own. This honest engagement with audiences has consequences for the language planners use. For example, seeing myself as a public intellectual (Jacoby 1987) who actively mediates among diverse planning audiences, I have tried to let the characters in this book speak their own rhetorics (which are often highly technical, passionately inflammatory, or politically opaque) and then to translate their words into the ordinary language of public discourse.

The fifth principle is that planners must be open to transformation through dialogue. They must be willing to expand the boundaries of their initial interpretive community or perhaps even to construct a new community. Planners must tell a story, and they must narrate it in a compelling and persuasive way, but they must also engage conflicting stories and be open to transformation through the encounter with them. This need to engage others in dialogue and to be open to transformation through that dialogue, presumes the existence of settings (or forums) that are appropriate for dialogue. If such settings do not exist, then planners should strive to create them.

Considered together, these principles suggest planners should strive, not to do good and be right and get things done, but to create, sustain, and participate in a public, democratic discourse that enables them (and others) to argue persuasively and coherently about contestable views of what is good, right, and feasible.

Interlude

Attending an Electric Utility Conference

I flew into Kansas City, Missouri, last night and now find myself in the new Marriott Hotel near the Country Club Plaza about forty-five blocks south of downtown. Not an electric power planner as such, I want to develop a better sense of how electric power insiders view the industry's current situation. I rise from bed, feeling unrested like I always do at conferences. New bed, new sounds. I shower, then take the elevator downstairs for the obligatory roll and cup of coffee. People, mostly older white men dressed in conservative suits, mill around waiting for the conference to begin. Brochures for this "1989 Utility Strategic Issues Forum," sponsored by the Electric Power Research Institute, say that it seeks to provide an opportunity for representatives from utilities, industry, government agencies, consulting firms, and EPRI to share perspectives and insights related to the future of the electricity business. "Where are the consumer and environmental groups?" I wonder. The keynote speaker proclaims that the electric power business has changed fundamentally and irrevocably, that utilities now face competition from many new players, and that utilities need a well-developed strategy if they want to win in this new environment. The next twenty or so speakers reinforce those opening remarks by talking of change, increasing competition, uncertainty, risk, the greenhouse effect, flexibility, spin-offs, mergers, acquisitions, and contingency planning. Acronyms abound: EPRI, NOIs, PURPA, FERC, IOUs, IPPs, NOPRs, NIMBYs, GNP, EPA, NUGs. "The electric power system," says one speaker, "is a machine." The ultimate question, says another, is "who controls it?" A third speaks of a "crisis of control." Still another shows a 35 mm slide of an electric power policy maker rowing a small boat, buffeted by a storm and surrounded by sharks.

3

The Modernist Institution and Rhetoric of Regulated Natural Monopoly

The meeting referred to in the prelude of this book occurred on February 11, 1991. The seven-member Illinois Commerce Commission (ICC) had convened to hear the counsels for opposing sides present their best arguments about whether the Commission should authorize the Commonwealth Edison Company (Com Ed, or Edison) to increase the rates that its customers pay for electric power. The price increase would compensate the company for having built three very expensive nuclear power plants: Byron 2 and Braidwood units 1 and 2. Held at the ICC's hearing chambers in Chicago, Illinois, the hearing was attended by the commissioners; the Commission staff; and attorneys representing Edison, the Cook County State's Attorney, the City of Chicago, the Illinois Office of Public Counsel, Business and Professional People for the Public Interest (BPI), Illinois Industrial Energy Consumers, the Governor's Office of Consumer Services, the Illinois Small Business Utility Advocate, and the Illinois Attorney General's Office. Other, unidentified people were also present. The handsome young man was Terry Barnich, recently appointed by Governor James Thompson to chair the Commission.

Why were these people meeting to decide whether Com Ed should be allowed to raise its prices? One good answer, I suggest, was to reenact the electric power industry's version of modernist planning and regulation in the United States. To clarify that modernist story, let me begin by outlining its plot and recognizing the figurative language that holds it together. Then we can focus on Edison, its plan to expand nuclear generation of electrical power, and the responses to that plan.

All this happens within the framework of the modernist story. By the end of this chapter, we should understand why the modernist story no longer can be told with confidence, at least in Illinois. In fact, we can glimpse how it is already being replaced by at least two other, conflicting stories, though both still get told in an institutional setting designed for the modernist tale.

A Regulated Natural Monopoly Has Certain Rights and Obligations

As can be inferred from the opening remarks of the Edison attorney, the modernist story begins with "basic economic theory." According to this story, the electric power industry exhibits sizable *economies of scale.* Large power plants, though more expensive to build, cost less for each additional unit of capacity; as generating capacity increases, average costs per unit of production decrease and the product becomes cheaper to produce. Accordingly one large firm or large plant can serve consumers at a lower average cost than can many smaller firms or plants. To take advantage of these economies of scale, a relatively large amount of money must be invested in plant and equipment (rather than labor or fuel) to produce a single dollar of revenue. That makes the electric power business *capital-intensive.* Whereas the average manufacturing facility requires about $0.9 of investment to generate a single dollar of revenue, the typical electric utility must spend nearly $4 on plant and equipment for each dollar of revenue. According to economic theory, an industry that displays such economies of scale and capital intensity is best served by one large firm. It is a *natural monopoly.* Edison is that monopoly for the northern one-third of Illinois, including the city of Chicago. In 1985, it had $4.1 invested for every dollar of gross revenue in order to provide electric service to about three million customers. These represented about eight million people living in a service area of about 11,525 square miles in northeastern Illinois (see fig. 3.1) (Moody's 1988).

Though Edison owns the electric power system and acts in some ways as a "natural monopoly," the company cannot act unilaterally. Liberal-democratic governments have consistently acted to protect ordinary citizens from abuses of monopoly power by regulating public utilities. Since 1913, the ICC has been required to review and approve any changes in Edison's rates and its plans for construction of new facilities before they can take effect.

Though economic theorists speak of "natural monopoly," historians suggest that the founders of the electric power industry labored hard to drive competitors out and then to

Edison is subject to many other forms of regulation as well. The Federal Energy Regulatory Commission (FERC) regulates interstate sales of electric power; the Security and Exchange Commission (SEC), the FERC, and the ICC regulate the financing of new capacity additions. The FERC, the ICC, the U.S. Environmental Protection Agency (EPA), the Nuclear Regulatory Commission (NRC), the Occupational Safety and Health Administration, and a wide variety of state and local agencies regulate Edison's operations with regard to service, environmental effects, and public health and safety. Furthermore Edison serves the City of Chicago and its residents under a franchise agreement that was initially adopted in 1907 and renewed in 1948. This franchise was scheduled to expire in December 1990, unless jointly renewed by Edison and the City.[24]

keep them out. Once firmly established, these monopolies were able to take advantage of scale economies. The claim that the industry *is* a natural monopoly now serves to distract attention away from technological changes that promise to transform power generation into a competitive market (see Kahn 1988).

The notion of a regulated natural monopoly has traditionally implied a *regulatory bargain* that imposes a set of rights and obligations on utilities like Commonwealth Edison. As shown in table 3.1, this traditional "bargain" involves four basic obligations and five basic rights (Phillips 1993, 118–20). Perhaps the most basic element in this "regulatory bargain" is that Edison is obliged to provide electric power services to any customer in its service area on demand. In return, the company has the right to charge "reasonable" rates. This bargain has profound implications for how Edison and other electric utilities have traditionally planned their investment in new facilities to generate power and how they have sought rate increases to pay for those facilities. Let us review those implications briefly, skimming over the surface of an extremely complicated process. Further details can be added later.[25]

The Modernist Approach to Forecasting Electric Power Demand and Planning Capacity Expansion

Experiencing economies of scale, the electric power industry achieved steadily decreasing costs from its birth in the 1880s through the late 1960s. Decreasing costs meant larger markets, which in turn meant more efficient power plants, leading to still lower costs and more growth. In conditions of decreasing costs, the task of electric power planning was quite simple. The utility would extrapolate past trends in the demand for electric power into the future and then identify the

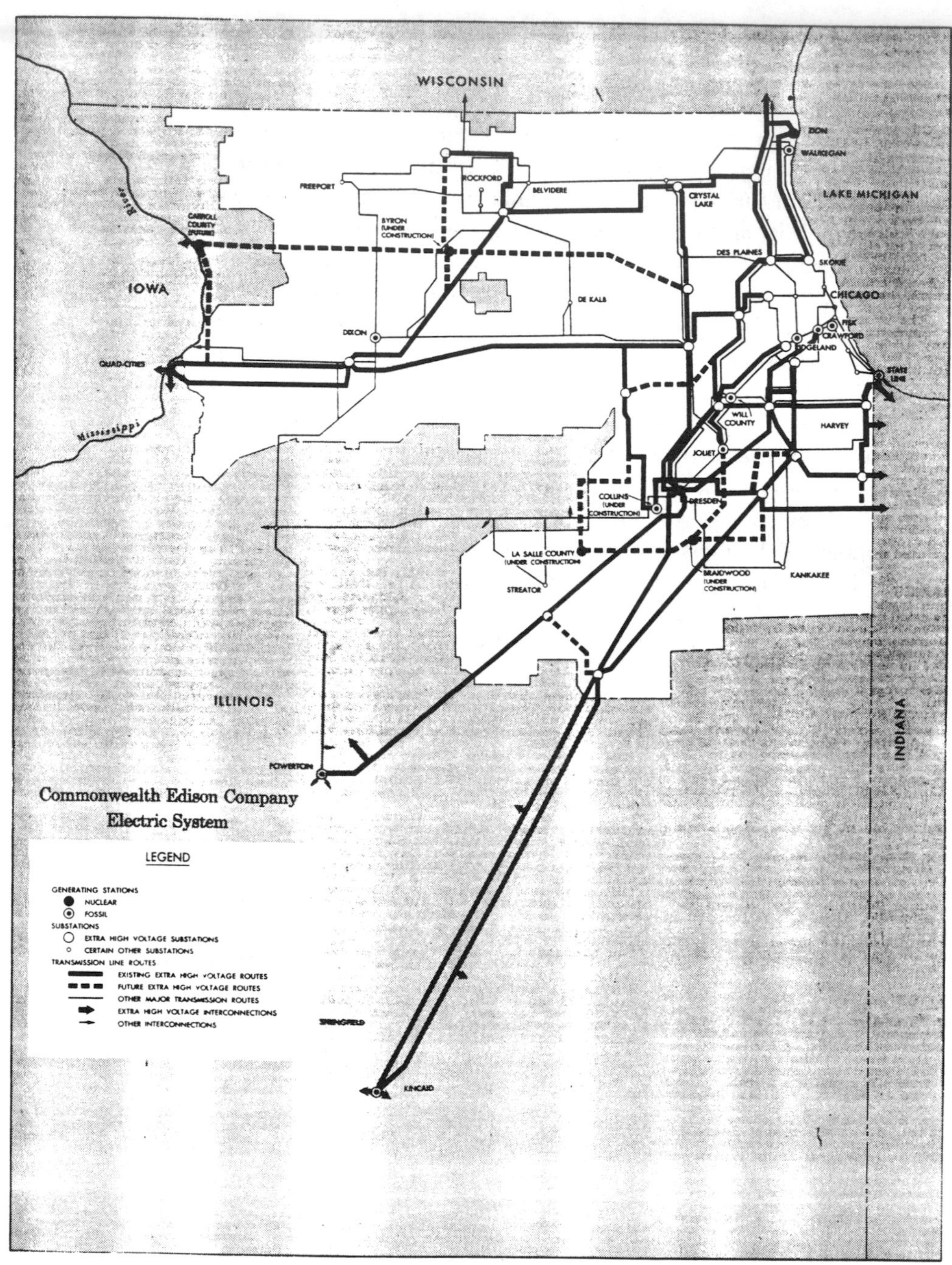

FIG. 3.1 **Commonwealth Edison Company's Electric System in 1977**

TABLE 3.1 **Traditional Rights and Obligations of a Regulated Natural Monopoly**

Obligations	Rights
• Edison must serve all who apply for service.	• Edison has the right to legal protection of its private property.
• Edison must render safe and adequate service (which implies instantaneous service on demand).	• Edison has the right to charge a reasonable price for (but not necessarily to earn a reasonable return on) its service.
• Edison must serve all customers on equal terms (within reason).	• Edison has the right to render service subject to reasonable rates and regulations.
• Edison must charge a "just and reasonable" price for the services it renders.	• When it furnishes adequate service at reasonable rates, Edison has the right of protection from competition from an enterprise offering the same service in the same area.
	• Edison has the right to condemn private property and take it for "public use" in return for payment of just compensation.

least costly but still reliable way to meet the forecasted growth in demand.

Consider, for example, a forecast Edison might have produced in 1972 based on total kilowatt-hour (kWh) sales from 1961 through that year.[26] According to a simple linear extrapolation of recent trends (see figure 3.2), Edison's sales would increase by 33 billion kWh (or about 5 percent per year) from 1972 through 1985. Since sales actually grew at an average annual rate of 7.8 percent from 1961 through 1972, this linear model would have underestimated future sales (Moody's 1967, 1972, 1977). An exponential model would have been more appropriate.

Reality, of course, is not quite so simple as even an exponentially based trend model would suggest. Since electric power must be generated when the customer wants it (when you flick on the light switch, for example), the electric utility must have generation *capacity* available to meet the maximum demand ("peak load") on its system—whenever it occurs. Obligated to serve all customers, it must anticipate future needs for electric power and

This claim of "obligation" makes the company's claim that it needs to build additional power plants more persuasive. Note, however, that the company could charge different

prices for different levels of service. At periods of peak demand it could—pursuant to prior contract—cut or reduce service to some customers and thereby reduce its peak capacity requirements. Projected "needs" for additional capacity are, in other words, intimately connected with demand in ways that simple statistical models cannot capture.

make whatever investments are necessary to meet those needs efficiently. It must provide enough generating capacity to meet these peak loads while still ensuring that sufficient capacity will be available in the event of a planned or unexpected outage. This is called having an adequate "reserve margin." So the utility must forecast peak load. A forecast based on simple linear extrapolation of recent trends would have shown that Edison's peak load was growing at a rate of roughly 5.3 percent per year and that it would have to add 8,000 megawatts (MW) of additional capacity between 1973 and 1985 to meet the load projected for the latter year (see figure 3.3). An exponential model would indicate that even more additional capacity would be needed.

The costs of even relatively small errors in forecasting can be quite large. Err on the low side, and the utility risks having insufficient power to meet demand. Blackouts might result. Err on the high side, and the utility risks having excess capacity. Its "capacity factor" and "load factor" would be low, and its costs would have to be spread over fewer kilowatt-hour sales. The per unit costs would go up. To make matters more complicated, a large power plant typically takes eight to twelve years to build, so electric utilities have to begin building plants well in advance of the demand.

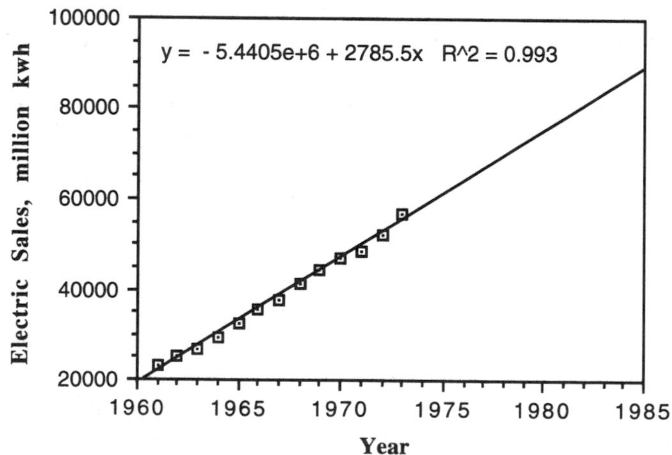

FIG. 3.2 **A Forecast of Edison Sales Based on Linear Trend Extrapolation**

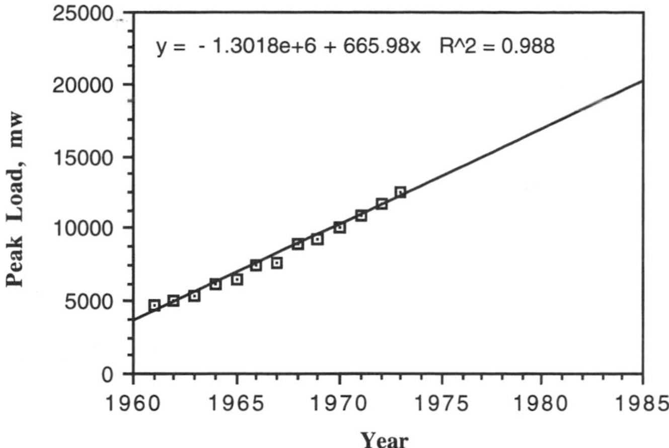

FIG. 3.3 **Trends Indicate Edison Needs 8,000 MW More Capacity by 1985**

The Modernist Approach to Electric Power Ratemaking

Regulatory commissions can only approve rates that are "just and reasonable." According to the modernist view of rate regulation, specification of "reasonable" rates is a balancing act: the Commission must *balance the interests* of Edison's owners with the interests of its customers. Owners seek a "fair" return on the capital they invest; customers seek rates that are affordable (Gormley 1983).

The notion that commissioners "balance" interests presumes that those interests are commensurable and can be reduced to constant dollars. What happens to concerns and issues that cannot be reduced to such a simple metric?

There are no substantive standards for determining whether this balancing act produces reasonable rates. Rather, the reasonableness is procedural.[27] The balancing hinges on the "due process" built into the ICC's deliberations. According to the modernist principles of rate making long-practiced in America, the Commission can best ensure due process by following a clear path of decision. Such decision making normally begins with the utility filing a request for a rate change. That often elicits quick briefs in reply from "intervenors," interested parties who support or contest the utility's claims. Presented with such a request, the Commission creates a file (a "docket") that documents the path, and it assigns the case to a hearing examiner. The examiner conducts "evidentiary hearings," where expert witnesses present detailed evidence that assesses whether the rate increase would be justified. Such expert testimony normally focuses on the utility's "revenue requirements." The examiner hears the direct testimony of these witnesses, their cross-examination by opposing counsels, and their rebuttal testimony. After reading the briefs and reply briefs and weighing the testimony, the hearing examiner writes a proposed order for the full Commission.

"Public participation hearings" are often held at this point, at least in Illinois. They have no direct influence on the examiner's findings of fact and law, and they have no significant bearing on the calculations embedded in the revenue requirements formula. But they can affect how individual commissioners weigh interests as they engage in their delicate balancing act.

After hearing oral arguments, the commissioners then adopt (by a majority vote) a formal order that accepts, amends, or revises the hearing examiner's proposed order. Any dissatisfied participant can appeal the order to a circuit court of the State of Illinois, then to the Illinois Appellate Court, and eventually to the Illinois Supreme Court. Once the order is adopted, the utility files its new rates with the Commission and begins billing customers. So long as due process is followed, the precise nature of that delicate balance of investor and consumer interests is left to the discretion of the Commission.

Though the reasonableness of rates hinges on due process, the procedures focus on certain key substantive issues. Most importantly, the Commission must determine the total revenue that the utility requires to provide its service (its "revenue requirement"), then translate that total into a rate structure for the utility's customers (Phillips 1993). Thus the Commission selects the year of record. This "test year" can be either historical or forecasted. Then it calculates the utility's revenue requirements for that year, compares test year earnings after taxes with allowed earnings, and adjusts rate levels so that expected revenues after taxes equal the permitted revenues.

The utility's revenue requirement can be determined by formula:

$$R = O + (V - D)r$$

where R = total revenue requirement
O = operating expenses
V = value of tangible and intangible property
D = accrued depreciation of tangible and intangible property
r = rate of return

Operating expenses include fuel (e.g., coal and natural gas) costs; regulation or rate case expenses; salaries, wages, and fringe benefits; advertising, promotion, and public relations expenses; contributions, donations, and dues; merchandizing and lobbying expenses; transactions among affiliated companies; amortized expenses; depreciation of property; and taxes. Operating expenses normally comprise the majority of electric utility revenues, averaging about 59 percent of total revenues. Depreciation and taxes average about 10 and 13 percent of total revenues respectively (Phillips 1993).

Controversial issues can be embedded in each category. Take fuel costs. In the mid-1970s, many state commissions adopted automatic fuel adjustment clauses, allowing utilities to charge customers for fossil fuel price increases without filing for a rate increase—and thus without having to go through a lengthy rate case. One consequence of such clauses is that utilities have little incentive to minimize fuel costs; another is that subsidiary companies (e.g., coal mining) owned by the utility might be able to earn extraordinary profits. Advertising expenses offer another example. Promotional advertising used to be considered an ordinary cost of doing business, but escalating costs of new generating capacity in the 1970s and 1980s have made such advertising controversial.

The *rate base* ($V - D$ in the formula) is the "value of a utility's property used and useful in the public service minus accrued depreciation" (Phillips 1993, 315). In other words, it is the accumulated capital cost of facilities purchased or installed to serve customers. These are the facilities on which the utility is allowed to earn a return. The elements of value to be included in the rate base are tangibles ("used and useful" land, buildings, and equipment), working capital, and property held for future use. Intangibles—including customer good will, customer contributions, and tax deferrals—are occasionally included in rate bases.

No clear-cut criteria existed in Illinois prior to 1985 with regard to whether a new facility should be considered "used and useful" for rate making purposes. Legal precedent indicated, however, that the Commission had been considering such factors as need, economic dispatch, economic reasonableness, enhanced system reliability, capacity used to serve ratepayers, use in carrying out the purposes of the utility, excess reserve margin, and providing low fuel costs (Illinois Supreme Court 1991).

Determining the value of utility property may seem a straightforward economic calculation, but it is not. In the case of regulated electric utilities, value depends on earnings, and utility earnings depend on the rates customers are charged. The rates are established by regulatory commissions. "Regulatory commissions are not finding the value of the property; they are making it," Phillips observes. "If rates are set high, value will be high; if low rates are set, value will be low"

(1993, 316–18). Even so, measurement of the value of utility property is crucial. It can be measured in terms of "original cost" (what the facility cost at the time it was built or purchased), "reproduction (or replacement) cost" (what the same or equivalent facility would cost if built or purchased at the present time), or "fair value" (a figure that falls somewhere between original and reproduction cost and is arrived at through Commission judgment). Regardless of the measure of value, the property's depreciation over time must be taken into account. Property that is used up in service is charged to operating costs; the remainder is included in the rate base. The service life of the property (its depreciation period) must be defined and then the depreciation charge distributed over the expected service life.

The *rate of return* is the amount of money earned by an electric utility, over and above operating costs, expressed as a percentage of the rate base. It includes interest on long-term debt (bonds), dividends and preferred stock, and earnings on common stock (including retained earnings). Investors expect to receive a "fair" return on the capital they invest. Therefore the rate of return is supposed to assure the utility's financial integrity, attract additional capital as needed, achieve earnings comparable to companies with comparable risks, and (perhaps) reward greater managerial efficiency. What constitutes a fair rate of return? It must attract capital but not exploit customers. Accordingly the courts have ruled that the concept of a fair rate of return represents a fairly broad "zone of reasonableness" (Phillips 1993, 375).

Since operating costs comprise roughly 80 percent of a typical electric utility's total revenue requirement, customer rates depend primarily on operating expenses. Even so, a reflective glance at the revenue requirements formula reveals that the company's earnings depend not on operating costs but on the product of the rate base and the rate of return. To increase earnings, a company must increase the value of its rate base, obtain a higher rate of return, or achieve a combination of the two. This creates, in the view of some analysts, an incentive for electric utilities to expand their rate bases beyond what would be economically optimal.[28]

The *rate structure* designates the specific rates or prices ("tariffs") that will yield the utility's required revenues. A "sound" rate structure generates the required revenues, apportions the costs of service among customers (avoiding "undue" or "unjust" discrimination among customers), and enhances economic efficiency. Such rate structures are based on "embedded cost-of-service" studies. "Cost of service" refers to the utility's costs incurred in serving each customer class (e.g., residential, small commercial and industrial, large commercial and industrial).

It includes a reasonable return on investment. The costs are typically divided into three categories:

- Demand costs vary in proportion to the customer's peak load (the peak number of kilowatts required by the customer). These apportion the utility's generating capacity, and costs of building it, to customers.
- Energy or commodity costs vary with the amount of energy each customer uses. These apportion the utility's energy and other operating costs.
- Customer costs vary in terms of the number of customers served. These allocate the utility's cost of metering, accounting, and billing its customers.

In broad outline, this is the modernist story about electric power planning and regulation in America. Year after year, in state after state, in case after case, utilities and consumers have stood at podiums pleading for regulatory commissioners to approve rates that are just and reasonable and in the public interest. Though they often differed over details, these characters told and retold, lived and relived a shared story with a simple plot: electric utilities should build more power plants to reduce operating costs and increase utility earnings and thereby reduce the price of electricity to the benefit of both consumers and the utility. Regulatory commissioners (in this case the ICC) have sat comfortably in quiet chambers, balancing the interests of electric utilities (in this case Edison) against those of consumers in a context of declining costs. They and the counsels who pleaded their cases have all spoken the figurative language that held their collective story together. They have spoken of natural monopoly, regulatory bargain, just and reasonable rates, due process, revenue requirements, the fair value of the rate base, and a fair rate of return.

The Illinois Commerce Commission stated its own view of this balancing act quite explicitly in a 1982 Edison rate decision:

> On first impression, it would appear that the commission is faced with two competing interests in its task of regulating; i.e., the ratepayer (consumer of the service) versus the shareholder (the provider of the service). It is critical . . . to recognize that while these interests must be balanced, the interests of both the ratepayers and shareholders are interdependent. The consumer needs the service; the provider of the service needs the consumer. With these principles in mind, the commission's obligation then is to ensure that efficient and effective service is provided to the ratepayer at a price which is fair and reasonable and sufficient to maintain the financial integrity of the utility. (ICC 1982b, 243)

The modernist story was told and retold with great confidence for many years. Beginning in the early 1970s, however, its characters—who were also its authors—found the story becoming harder and harder to repeat with conviction. It became harder and harder to make the simple outlines of the modernist plot fit the convoluted events of their regulatory world. As chapter 4 documents, the utilities' revenue requirement began to rise, and the ICC began having great difficulty balancing producer and consumer interests in its Edison rate case decisions. Edison began complaining that it had not been allowed to receive a fair return on its investment, whereas consumers routinely expressed bitter outrage at being exposed to unjustified, unreasonable, and unaffordable rate increases.

Time and again, the smartly dressed men and women at the front of the room found themselves frowning as they faced rooms full of increasingly contentious and confusing counsels.

Interlude

On a Riverbank West of Chicago

We lie on the sand together, the waters of the mighty Mississippi gently lapping near our feet. A warm, late-summer day. The sun filters through the canopy of trees, warming our bodies. The trees surrounding us sway in the lazy breeze. Crows caw to one another, complaining about our presence or about some other offense that we cannot fathom. As I lie there, eyes closed, she holds some item of food close to my face, asking me to smell and then taste it. "What is it?" she asks. Guessing, then opening my eyes, I discover I had guessed wrong. We laugh. Totally alive, I can feel the hairs on my skin dance with the trees. Gradually, though, I begin to notice the sound of two fishermen in a motorboat out on the river. Coming our way. Across the broad muddy stream, about half a mile south of us, we can see a large electric power plant. "The Quad Cities nuclear power plant," she says. "It's owned by Commonwealth Edison. They built it about twenty-five years ago." The smell of dead fish. Extending her hand, she smiles. "Let's find some other place."

4

Commonwealth Edison's Ambitious Nuclear Power Expansion Plan, 1973–1986

A Plan for Growth

In 1956, Edison obtained the ICC's permission to build Dresden 1, the nation's first full-scale, privately financed nuclear power plant. Almost one decade later, with its forecasts showing the demand for electric power growing at 7 percent per year, and believing that nuclear power would be the best way of providing adequate and reliable power in the future, Edison announced a major expansion of what has become the nation's largest program to construct nuclear power plants. Between 1965 and 1973, the company announced its intention to build twelve or more large nuclear units at its Dresden, Quad Cities, Zion, LaSalle, Byron, and Braidwood stations. More specifically, in 1970, 1972, and 1973, Edison applied to the ICC for "certificates of public convenience and necessity" to authorize and direct construction of six nuclear generating units totalling approximately 6,600 MW at its LaSalle, Byron, and Braidwood stations. According to material submitted with those applications, the units would be completed by 1980 at a cost of $2.5 billion (ICC 1980a; ICC 1985b; BPI and CUB 1989b). They would extend the modernist story of a regulated natural monopoly building larger power plants that increased earnings and reduced consumer rates.

This was an ambitious plan. As time passed, it also proved to be highly controversial and difficult to carry out completely or on time. The plan assumed that the economic stability of the 1960s would continue. It also assumed that nuclear power plants, which heretofore had been designed as 300 MW units, could easily be modified to operate

TABLE 4.1 **Edison's Nuclear Expansion Plan Finished Far over Budget and behind Schedule. Why?**

Unit	Planned	Actual
LaSalle 1	1975	1982
LaSalle 2	1976 ($567M)[b]	1984 ($2,490M)[b]
Byron 1	1979 ($437M est)	1985 ($4,180M)[c]
Byron 2	1980 ($546M)	1987 ($1,884M)
Braidwood 1	1979 ($506M)	1988 ($3,268M)[c]
Braidwood 2	1980 ($450M)	1988 ($1,863M)
Carroll Co. 1 & 2	1992 and 1993	Deferred
Total	($2,506M)	($13,685M)[a]

[a] *Sources:* ICC, 1980a; Com Ed 1985; BPI and CUB 1989b.
[b] Includes LaSalle 1.
[c] Includes the cost of facilities commonly shared with the second units at each station.

equally efficiently at the 1,000 MW scale. Consequently this plan was shaken by the energy price shocks of the 1970s, by price inflation, by the 1979 nuclear accident at Three Mile Island. Moreover, it was assaulted by the rise of consumer, environmental, and antinuclear movements.

Edison's plants took longer to build, and they cost much more than expected. The last of the six nuclear units was not completed until 1988, and the final construction cost turned out to be approximately $13.7 billion (see table 4.1). Thus Edison's nuclear plan was completed eight years behind schedule and $11.2 billion over budget!

To pay for its increasingly expensive plants, the company had to apply for a series of large rate increases. As the price of electricity increased, the rate of growth in demand for electric power declined. Edison's capacity and load factors decreased, while its reserve margin increased. Faced with steadily increasing electricity prices, consumers looked for alternative ways of obtaining needed power and energy services.

What happened? Was the original plan itself bad? Were its forecasts of demand and construction cost and other elements so faulty technically that one should think of Edison's experience as just another "great planning disaster" (Hall 1980)? Was Edison's a good plan that became mired in the morass of Illinois politics?

I want to suggest that Edison's experience marks a transition in planning: a decisive movement away from modernist planning and toward other, "postmodernist" forms of planning. Speaking in the normal discourse of regulated natural monopoly, but experiencing events that undermined that institution's rationale, Edison and consumer groups started telling competing stories about why Edison's plan had gone awry and about what should be done as a result. Yet the utility continued to speak the modernist story, blaming the unplanned twists in its plot on events beyond its control and on politically motivated consumer intervenors. Consumer groups, on the other hand, also extended parts of the modernist account by blaming the twists in the plot on incompetent utility managers. By the consumer group account, these managers had sought to increase profits by expanding the building plants that were not needed. At the same time, however, the consumer groups joined the struggle to invent a new language and a new institutional structure for electric utility planning. And the commissioners—ah, the commissioners: they struggled to "balance the interests" of characters who were beginning to bewilder them by telling very different stories about planning and ratemaking in Illinois, stories that implied fundamental changes in the very notion of balancing interests through regulatory action.

Implementing Edison's Nuclear Power Expansion Plan

In October of 1970, the Commonwealth Edison Company asked the ICC to authorize construction of a 2,200 megawatt electric power–generating station in LaSalle County, Illinois. One unit would go on line in 1975 and the other in 1976, at a total cost of $567 million. Edison presented evidence during hearings that peak demand in the company's service area would reach 17,850 MW by 1977 and that existing generating capacity was not sufficient to meet that demand. So, Edison proposed to build a new generating facility. Furthermore, the company argued for building a nuclear plant, rather than a fossil fuel plant, on the grounds that nuclear power would be less expensive and less harmful to the environment. Edison's application attracted no significant opposition (ICC 1971).

Applications for certificates of convenience and necessity for additional stations quickly followed. In 1972, Edison asked the ICC to authorize construction of two 1,120 MW units at its new Byron station. In the following year, Edison asked to be allowed to build two 1,100 MW units at Braidwood, both to be completed by late 1980. In every case, Edison presented forecasts which indicated that growth in peak demand necessitated another plant.

Much to Edison's surprise, and probably to its dismay, growth in the company's electric sales and peak load slowed considerably in the years immediately following the ICC's approval of the Braidwood units (see fig. 4.1). Whereas sales had grown at an average annual rate of 7.8 per-

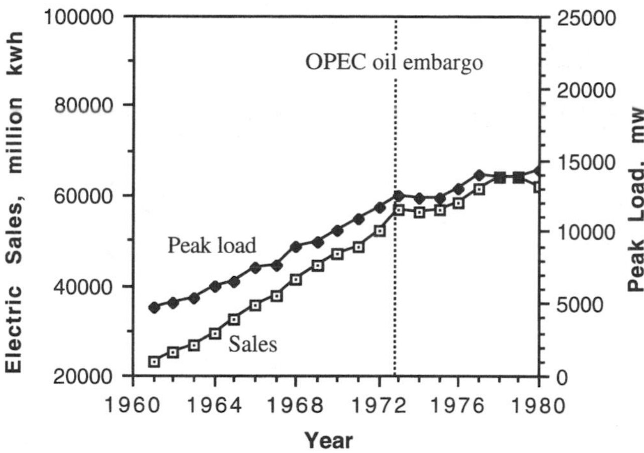

FIG. 4.1 **Edison's Sales and Peak Load Growth Weaken after 1972**
Source: Moody's (1967, 1972, 1977, 1982).

cent from 1961 through 1973, in 1974 they actually declined by 1.5 percent. Peak load declined by the same amount. Convinced that these declines were only temporary, however, the company continued its expansion program. This initiated a long-running debate about whether the new plants were needed and who should pay for them.

The first request for a rate increase generated little controversy. Only a few scattered organizations intervened after the company applied for a $154 million (12.2 percent) increase in May 1973, and the Commission's approval of a $134.7 million increase (10.7 percent) in April of the following year did not seem too worrisome (ICC 1974). In late 1974, however, Edison asked for another $241 million (or 15.6 percent) increase. Claiming that it was experiencing a sharp decline in earnings, Edison also asked the Commission to approve an interim rate increase of 7.3 percent to take effect immediately.

If rates were not increased, the company said, it would not be able to attract the new capital needed to meet the company's forecasted construction requirements. Though the rate increase was opposed by a large number of intervenors, mostly labor unions, the Commission granted the company interim "rate relief." Then, in August 1975, it approved a permanent increase of $207 million (or 12.7 percent).

Edison's sales and peak load increased only slightly in 1975, but the company remained optimistic that its pre-1974 growth rates would return. In late 1976, it applied for another rate increase, this time for $263 million (or 14.5 percent). Price inflation was driving up the cost of

Public Health Risks and the Economics of Nuclear Power

Though the debate over electric power rate increases might appear to be a matter of straightforward economics, thoughtful reflection reveals that the debate is—at least with regard to nuclear power—deeply political as well

The nuclear power industry grew out of atomic weapons research relating to World War II and the subsequent Cold War. Its earliest years were shrouded in secrecy and military ventures. A breakthrough came in 1953, however, when President Eisenhower announced a national effort to promote "Atoms for Peace." Privately owned electric utilities initially hesitated to build nuclear plants, but passage of the Atomic Energy Act of 1954 helped alleviate many of their concerns. With little public debate, the federal government committed itself to massive subsidization of nuclear power. The chairman of the Atomic Energy Commission (AEC) proclaimed that "It is not too much to expect that our children will enjoy electrical energy *too cheap to meter*" (Hilgartner, Bell, and O'Connor 1982, 44). Shortly thereafter, the nation's first commercial reactor went into operation at Shippingport, Pennsylvania. The federal government then initiated a five-year Power Reactor Development Program that heavily subsidized private research and development costs for prototypical nuclear plants. Commonwealth Edison built its 200 MW Dresden 1 unit as part of this program. In 1957, a Brookhaven Institute report (WASH–740) indicated that a major nuclear accident was highly unlikely, but that the consequences of such an accident could be devastating. Persuaded that private utilities would not build nuclear power plants in the face of such financial risks, Congress enacted the Price-Anderson Act that same year, setting a ceiling of $570 million on private liability in the event of a serious nuclear accident.

The private nuclear industry grew slowly until 1963, when the General Electric Company offered to guarantee the costs and completion dates of new plants. Thirteen plants (including Edison's Dresden 2 and 3 and Quad Cities 1 units) were built during the "Great Bandwagon Market" that followed over the next few years. Orders for new nuclear units shrunk later in the sixties, and licensing came to a halt for eighteen months while the AEC responded to the Supreme Court's 1971 Calvert Cliffs decision. Proponents revived interest in the early seventies, despite the lack of actual cost experience for units of the size then being built, and electric utilities ordered another 114 nuclear units between 1970 and 1973. Seeking to reassure the public, the AEC published WASH–1400 (the "Rasmussen Report") in 1975. It calculated that the "maximum credible" accident would occur only once every ten million operating years, but that the worst possible accident would produce 3,300 fatalities, $14 billion in property damage, and evacuation of 290 square miles.

Proponents interpreted the OPEC oil embargo of 1973 and subsequent price increases as evidence that nuclear power was the ideal energy source. Nevertheless, citizen groups at various locations around the country organized to oppose nuclear power, most notably at Seabrook in New Hampshire. These groups believed that nuclear units were being built without adequate safety precautions and without adequate concern for the long-term disposal of nuclear wastes. In 1974, Ralph Nader organized a national antinuclear "Critical Mass" conference in Washington, D.C. to inspire a fight for utility rate reform.

The Rasmussen Report's calculations notwithstanding, a serious nuclear accident nearly happened at TMI in 1979. Roughly 144,000 people evacuated the area, over $1 billion was spent on cleaning up the damaged reactor, and shareholders lost roughly $800 million in the eight years after the accident (Houts, Cleary, and Hu 1988). The accident led to creation of the Kemeny Commission and caused the NRC to adopt new safety regulations. Public confidence in nuclear power fell. (For details, see Bupp and Derian 1981; Pringle and Spigelman 1981; Hilgartner, Bell, and O'Connor 1982; the U.S. Office of Technology Assessment 1984; Perrow 1984; Campbell 1988; Houts, Cleary, and Hu 1988; Weart 1988; and Morone and Woodhouse 1989.)

building its new plants, Edison said, and it needed rate relief in order to attract additional capital. The Commission granted the company a $151 million increase (7.6 percent) (ICC 1977).

Sales and load growth continued to improve in 1976, rising by 5.3 percent and 7.9 percent respectively. In 1977, however, peak load actually declined by 1.5 percent, even though sales grew by 4.2 percent. In early January 1978, the company came back for another rate increase, this time for $125 million, or 5.6 percent. Price inflation had consumed some of its authorized earnings, the company argued, and the Commission should allow it to compensate for inflation by increasing its rates. Thirty-two intervenors took part in the proceedings, including the Illinois Public Action Council, Citizens for a Better Environment, the City of Chicago, and the Cook County State's Attorney. These intervenors objected to the proposed increase largely on the grounds that Edison's management had failed to keep costs down and had embarked on an overly ambitious and largely unnecessary program of construction. In essence, they argued that Edison "needed" a higher rate of return only to attract more capital to build more generating plants that were not needed. The Commission decided to grant a $75 million increase, but it also indicated that "a formal investigation of Edison's construction program should be commenced" (ICC 1978, 51917).

The ICC began its investigation of Edison's construction program, intending to compare the economic consequences of completing the Byron and Braidwood plants as scheduled to the consequences of (1) delaying both plants for one year or (2) delaying Bryon for one year and Braidwood for three (ICC 1980a). At that time, Edison projected that its peak load would grow at an average of 5.1 percent annually over the next ten years and that its Byron and Braidwood plants would be completed by 1982, even though its actual load growth had averaged 2.4 percent per year over the preceding five years (ICC 1980a).

Shortly after the ICC initiated its investigation, however, the nuclear power industry (indeed the nation as a whole) was shocked by the nuclear accident on March 28, 1979, at the Three Mile Island (TMI) plant in Pennsylvania. The industry quickly recuperated enough to interpret the accident as proof that the system had worked: an accident had occurred, but the safety procedures had prevented disaster, and no one had been killed. Opponents, on the other hand, interpreted TMI as proof that the system had failed: despite years of industry proclamations of impossibility, a nuclear reactor had suffered a near meltdown (Ford 1981).

The accident at Three Mile Island led the Nuclear Regulatory Commission (NRC) to suspend for six months licensing of all nuclear plants currently under construction, including Edison's. Citing the NRC's licensing delay, Com Ed announced in mid-1979 that the Byron and Braidwood stations would not be finished until 1983. The ICC instructed Edison to analyze the effect of further delays in completing the plants under construction. In May 1980, Edison announced that its LaSalle plant would not be finished until mid-1982, the two Byron units would not be completed until 1983 and 1984, and the two Braidwood units would be finished in 1985 and 1986. The company attributed the delays primarily to the need to modify plant designs in response to the NRC's post–Three Mile Island regulations. Edison also said that the increasing cost of the plants had forced the company to borrow more money than planned, at a time when money itself had become far more expensive. In any event, the company concluded that substantial penalties would be incurred for delaying construction even further, as opponents wanted. It maintained that the penalties associated with terminating Braidwood "were so great that no further study of such alternatives was necessary" (ICC 1980a, 7). Taking these factors into consideration, as well as others introduced by Commission staff and by intervenors, the Commission concluded that

> the incremental engineering economic studies conducted by Edison and the full revenue requirements studies conducted by the Commission's Staff demonstrate that substantial economic benefits will accrue to the ratepayers and the Company by completion of the Braidwood Station in as timely a manner as possible. This is true even though completion of the plants currently under construction will increase Edison's reserve margins. . . . The evidence . . . demonstrates that a forecasted increase in reserve margins does not indicate that delay in construction is economically advantageous when new generating units are approximately half constructed and their completion and use will displace the use of higher cost fuel. . . . The evidence further demonstrates that the delays . . . are introduced by many factors beyond the control of this Commission. It appears that the most recent delays announced by the Company are due to licensing and constuction constraints the Company is experiencing. The Commission is of the opinion that Edison has a duty to its ratepayers to complete the generating units under construction in as timely a manner as the construction and licensing constraints allow. (ICC 1980a, 11–12)

Here the ICC tries to balance the interests of producers and consumers. Note, however, that consumer groups had begun telling a story whose plot turned on profit-hungry utility managers building nuclear units that were not safe, needed, or wanted (by consumers). From this point of view, the ICC was in league with Edison. Either the composition of the Commission or the institutional structure of regulated natural monopoly would have to be changed. Or the consumer groups would have to keep swallowing bitter pills of policy.

So construction of the Byron and Braidwood plants continued, but the ICC also urged Edison to reconsider the need for two other new plants, one a coal-fired facility and the other a nuclear station proposed for Carroll County. At that point, the two Carroll County nuclear units were scheduled for completion by 1993, and $16 million had already been spent on the Carroll County plant, even though Edison had not yet obtained from the ICC the needed certificate of convenience and necessity (ICC 1980a).

While the ICC was investigating Edison's construction program, the company filed (in April 1979) a request for a $452 million general rate increase. Forty-one intervenors—including seventeen industrial firms and ten community-based consumer groups (such as Community Action for Fair Utility Practices and the Labor Coalition on Public Utilities)—took part in the case. Upon reviewing Edison's application, the

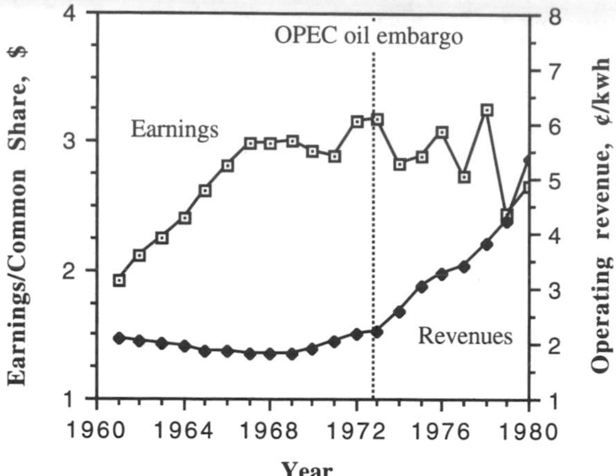

FIG. 4.2 **Edison's "Price" Increases after 1972, Earnings Decline**

Source: Moody's (1967, 1972, 1977, 1982).

ICC adopted 1979 as the test year, determined that the fair value of the company's rate base was $6,024.5 million, decided that a reasonable rate of return on the company's fair value rate base was 8.372 percent, and calculated that the company's operating revenues under current rates totaled $2,645.4 million. This would have given Edison an operating income of $324.3 million. Determining that $324 million did not constitute a fair return, the Commission allowed Edison to earn an additional $389.6 million. This 14.4 percent increase, adopted in February 1980, gave the company roughly 86 percent of what it had originally sought (1980b).

Six months after receiving this 14 percent increase, Edison asked the Commission to approve another 19.7 percent increase. This time, at least thirty-five intervenors (plus Edison and Commission staff) took part. The Commission held at least twenty-five evidentiary hearings and sixteen public hearings. Upon reviewing Edison's application, the ICC adopted the twelve-month period ending June 30, 1981, as the test year. It determined that the fair value of the company's rate base was $7,577.5 million. (This new figure included 15 percent of the company's current "construction work in progress" [CWIP] balance, or $729.6 million.) Deciding that a "reasonable" rate of return on the company's fair value rate base was 8.52 percent, it calculated that the company's operating revenues under current rates totaled $3,469.7 million. The result was an operating income of $411.5 million. Determining that $412 million did not constitute a fair return, the Commission then

Construction Work in Progress (CWIP)

The "fair value" of Edison's rate base included $343.5 million for facilities that were not yet "used and useful." This amount represented 10 percent of the company's current construction work in progress (CWIP) balance. Whether and how to incorporate into the rate base the cost of new facilities that were not yet used and useful had become a very controversial issue in the late 1970s, primarily because new generating units were so expensive relative to the net book cost of the firm's existing assets.

Edison and other utilities had long been allowed to include an Allowance for Funds Used During Construction (AFUDC); that is, the interest payments for financing new facilities. AFUDC, however, had at least one major drawback: it was only paper earnings, not cash. Consequently utilities could not use AFUDC to pay interest to bondholders or dividends to shareholders. Thus, utilities with large AFUDC tended to have lower bond ratings and higher capital costs. AFUDC accounts became very large in the seventies, accounting for more than 50 percent of the industry's return on common equity in 1980. (In Edison's case, AFUDC accounted for 94 percent of earnings!)

As electric utilities began investing in extremely expensive power plants, regulatory commissions like the ICC began to worry that the consequent rate increases might cause "rate shock" to consumers, local economies, and utilities. Consumers' bills would shoot upward, the demand for electric power might decline precipitously, the economy of the utility's service area might deteriorate, and utilities might enter a "death spiral." So commissions began exploring alternative ways to spread the rate increases over time. One was to phase in the rate increases associated with completed plants. Another was to include CWIP in the utility's rate base.

CWIP is a utility bookkeeping account that accumulates all costs, including AFUDC, associated with building new utility facilities until they become operative and are included in the rate base as used and useful. The inclusion of CWIP in the rate base became quite controversial. Utilities and some regulators argued that it avoided "rate shock" while also permitting a utility to earn a return on the capital invested in expensive new facilities. Consumer groups and others argued that CWIP should not be included in the rate base because the relevant facility had not yet become used and useful and because the expenses associated with building that facility had not yet been shown to be prudently incurred or reasonable.

allowed Edison to earn an additional $503.6 million. This further 14.5 percent increase, adopted in July 1981, gave the company roughly 74 percent of what it had sought in its second proposal.

The ICC used the company's proposed depreciation rate of 4.05 percent for nuclear plants. The company indicated that its current reserve margin was 12.4 percent, forecasted this to rise to 25.1 percent in 1986 (at completion of Braidwood), and predicted a subsequent decline to 15.9 percent by 1990. Seeking to ensure that Edison's reserve margin was at an "optimum," the Commission concluded that "there is no evidence in this case to support an adjustment to rate base due to 'excess capacity'" (ICC 1981, 521). The ICC also noted, however, that Edison's construction program and future generating capacity were being "carefully monitored" to consider if and when an adjustment should be made in the future.

One might wonder how the Commission balanced interests with regard to the disposal of high-level nuclear wastes. Edison had been storing those wastes "on-site" at each nuclear plant, but the U.S. Department of Energy had contracted to take them off Edison's hands beginning in 1997. Unfortunately, the DOE was having increasing difficulty finding an acceptable permanent disposal site. To pay for that disposal, the ICC had in 1980 instructed Edison to charge consumers one mill (a tenth of a cent) per kilowatt hour of nuclear generation. The 1981 order increased that to two mills per kilowatt hour. For details about planning for the disposal of nuclear wastes, see Carter 1987.

Construction of the LaSalle and Byron and Braidwood plants continued, and another year passed. In January 1982, the company applied for another rate increase, this time for $805 million (or 19.4 percent). At least thirty-seven intervenors took part, including the Attorney General of the State of Illinois. Edison stressed the need for half of that increase to take effect immediately, lest the company become unable to attract the additional capital needed to sustain its construction program. The Commission staff worried about that program, and suggested that "the 'traditional model' used in rate case proceedings for establishing an appropriate revenue requirement has lost much of its applicability . . . due to Edison's construction program and the current state of financial markets" (ICC 1982a, 69). Nevertheless the Commission continued in its next proceeding to rely on the traditional approach to ratemaking; and in December 1982, it approved another $660.7 million (17.1 percent) rate increase. Roughly 82 percent of what the company orginally asked for, this rate increase incorporated the effects of including LaSalle 1 (which had been finished in October) and $525 million of CWIP in the rate base. Five more units were still under construction, however, and the cost of building those plants worried the commissioners deeply. According to data presented at the time, those five plants would be finished by October 1986 at a total cost of $6.5 billion, an amount roughly equal to the net worth of all the utility's existing plants and working capital (ICC 1982b)!

Struck by the huge price tag, the Commission sought detailed testimony about the costs of delaying or canceling Braidwood or possibly Byron as well. The testimony raised at least four complicated issues for the Commission. The Attorney General, Business and Professional People for the Public Interest (BPI), and others questioned the usefulness of attempting to compare alternate courses of action in terms of their respective "net present value" revenue requirements. Many of those same witnesses also challenged the accuracy of data used in those studies. In particular, they argued that Edison's existing nuclear plants might have unanticipated modification and outage costs; that Edison had overestimated the capacity factors for plants under construction; that Edison had overestimated its future load growth; that Edison's existing fossil fuel contracts had locked the company into unnecessary expenses; that Edison's studies had overestimated the cost of replacement capacity; and that the company had seriously underestimated the potential benefits of load management and energy conservation. Their third criticism targeted the inequities associated with asking future generations to bear costs imposed by the present one. Many of these witnesses also doubted Edison's ability to finance its construction program. One of the intervenors, BPI, asked the Commission to investigate whether completing Byron 2 and the Braidwood plant would be economically superior to conservation alternatives. After considering the evidence, briefs, and arguments presented in the case, the Commission rejected most of the arguments for delaying or canceling Braidwood or any of the other plants. It concluded that

> it is in the public interest to complete the LaSalle, Byron, and Braidwood stations in as timely and economic a manner as good management practice permits and hereby directs Edison to complete its present construction accordingly. . . . These [engineering economic] studies, together with all the evidence, establish that the benefits to be derived from the completion of construction far outweigh the costs, and therefore, it is in the public interest to complete Edison's current construction program in as timely and economic a manner as possible.

If consumer groups were upset at the Commission after its 1981 decision, they had reason to be even more upset now. Once again, the ICC had instructed Edison to continue building plants that were neither safe, needed, nor

wanted (by the consumers). But at least the Commission had given BPI and others an opportunity to explore the costs and benefits of energy conservation.

The commission further concludes that . . . such a decison can be reasonably and properly made at this time and need not be deferred to determine if additional studies would show whether defined conservation alternatives and the costs associated therewith are economically superior to the completion of Byron Unit 2 and the Braidwood station. Given this conclusion, the pending motion by BPI in this docket should be denied.

However, the commission is further of the opinion that, while such conservation alternatives may be speculative as to specificity and cost, this should not act as a bar to investigation of such conservation proposals. Business for the Public Interest, as well as other concerned parties, are encouraged to file a proposal describing such conservation measures and the costs associated therewith so that the commission may provide a forum for an expeditious investigation. (ICC 1982b, 263–264)

Nine months later, in October 1983, Edison asked the Commission for another rate increase, this time of $462.2 million. Upon reviewing Edison's application, the ICC adopted 1984 as the test year. It determined that the company's rate base was $6,633.8 million. This determination included LaSalle 2, which would go "on-line" in June 1984, so the Commission explicitly chose not to calculate the current or fair value. The ICC decided that a fair and reasonable rate of return on the company's rate base was 12.82 percent, and it calculated that the company's operating revenues under current rates totaled $4,264.3 million. This produced an operating income of $702.9 million. Finding that $703 million did not constitute a fair return, the Commission allowed Edison to earn an additional $282.5 million. Adopted in July 1984, this 6.6 percent increase gave the company roughly 61 percent of what it had originally sought. But the Commission also expressed "concern" about the possiblity of "rate shock" to Edison's customers when the Byron and Braidwood plants were finished. Accordingly the commissioners indicated that they intended to consider how those plants could be phased into the rate base (ICC 1984).

Rate increase piled upon rate increase, steadily increasing revenues and net income for an already bloated electric utility.[29] The prospect loomed of even larger increases in the next few years. All this was occurring in a context of increasing unemployment and price inflation. At least that's what consumer and community groups saw in 1983. Partly in response to that sense of economic and political powerlessness, the voters of the city of Chicago had elected the progressive, reform-minded Harold Washington mayor in May of that year (Bennett 1988). Also responding to voter anger at the higher rates, Governor Thompson appointed two new, and more consumer-oriented, commissioners to the ICC. Margaret "Meg" Bushnell, a vocal consumer advocate who lived near the Byron plant, took her seat in November; and Susan Stone took hers just three months later. How would the new commissioners "balance the interests?"

Nineteen eighty-four began with what came to be known as "Black January" in nuclear power circles. The nuclear power industry had already been experiencing major difficulties nationwide. Utilities had ordered only fifteen new nuclear plants after 1974 and none after 1978. They had canceled orders for over one hundred nuclear units between 1974 and 1982 at a total cost of nearly $10 billion (Ittellag and Pavel 1985). In mid-1983, the Washington Public Power Supply System (WPPSS) defaulted on $2.25 billion in debt for four abandoned nuclear plants; and by the start of 1984 roughly one-half of the forty-eight nuclear plants then under construction were candidates for abandonment due to declining demand and soaring costs (Rudolph and Ridley 1986). Then in January, the owners of the Marble Hill and Zimmer plants in Indiana and Ohio respectively canceled those plants, even though $2.7 billion had been invested in the first and $1.9 billion in the second. Worse, the Zimmer plant was allegedly more than 95 percent complete (Ittellag and Pavel 1985). Not many months later, a front-page article in *Forbes* magazine announced that "The failure of the U.S. nuclear power program ranks as the largest managerial disaster in business history, a disaster on a monumental scale" (Cook 1985, cover page). The nuclear power industry seemed to be collapsing.[30]

In this context, the NRC's Atomic Safety and Licensing Board (ASLB) stunned Edison by denying its application for a low-power license for Byron 1. The main ground was that several of Edison's contractors had seriously deficient programs for assuring the quality of their quality-control inspectors. The ASLB feared that this might mean inadequate and unsafe construction. Shortly thereafter, the ICC hired the Arthur D. Little Company to audit that unit's construction costs. In October, the NRC decided that the licensing denial had been in error and approved the operating license (ICC 1985b).

That same month, a new consumer advocacy organization appeared. Soon known as CUB, the Citizens Utility Board was to represent residential and small business consumers in utility rate cases. Though created by the State Legislature, it was not a governmental agency. CUB was funded by voluntary contributions from its members, and the dues

NRC Licensing Procedures

The Atomic Energy Act instructs the NRC to issue licenses for construction and operation of nuclear power plants and to inspect plants and to enforce the NRC's licensing requirements. As of 1984, this licensing process had seven key steps (U.S. Office of Technology Assessment 1984):

- **The utility prepares and submits an application for a construction permit, including a Preliminary Safety Analysis Report (PSAR) and an Environmental Report (ER).**
- **The NRC dockets the application.**
- **The Office of Nuclear Reactor Regulation compares the application with the NRC's Standard Review Plan and prepares a Safety Evaluation Report (SER). An Advisory Committee on Reactor Safeguards (ACRS) reviews and comments on the application.**
- **The Atomic Safety and Licensing Board (ASLB) holds hearings on the application.**
- **The NRC grants a Limited Work Authorization.**
- **The application might be reviewed by the Atomic Safety and Licensing Appeal Board (ALAB).**
- **The utility applies for an operating license. In doing so, it must follow the same steps as for a construction license. The plant must also undergo extensive testing before going into commercial operation. No operating license may be issued without the utility having first prepared an acceptable Emergency Evacuation Plan, and the utility must reach agreements with state and local agencies on public notification, evacuation, and other procedures.[31]**

payers elected its twenty-two directors. Within months, its membership had soared to more than 120,000 people (CUB 1984–86).

There were strong signs that further—perhaps much larger—rate increases would soon be forthcoming. Byron 1 had suffered a licensing delay. Byron 2 and Braidwood limped along with continuing delays and cost overruns. The Commission decided to grant Edison a substantial rate increase for LaSalle 2. Work continued on Braidwood, and the ICC did not compel Edison to pursue conservation aggressively.

These signals alarmed and energized consumer groups. They continued to oppose Edison in ICC proceedings, but they also sought changes in the law that had governed the ICC's decision process for decades. And, in the turn of events most important for this book, they stimulated Chicago Mayor Washington's new administration to begin exploring alternatives to Commonwealth Edison's electric power system. (For details about Chicago's exploration, turn to chapters 7 through 9.)

Motivated by consumer hostility to Edison's expansion plan and by the ICC's relentless willingness to approve the rate increases that Edison required, the Illinois General Assembly began considering changes to the existing Public Utilities Act in April 1985. Guided by testimony and written comments provided in previous months, and pressed to reauthorize the act before it expired in December, the assembly's Joint Committee on Public Utility Regulation proposed several specific actions. It proposed to (1) strengthen the Commission's regulatory powers, (2) create a new Office of Public Counsel to represent utility customers in rate cases, and (3) promote "least-cost electric utility planning." It also proposed to (4) facilitate investment in energy conservation, (5) require the ICC to audit construction costs before allowing new plants into a utility's rate base, and (6) prohibit the ICC from including the cost of "excess" electric generating capacity in a utility's rates (Joint Committee on Public Utility Regulation 1985). Electric generating capacity that exceeded actual peak demand plus a 25 percent reserve margin would be deemed "excess," and the Commission would be required to exclude from consumer rates the common equity return on it. James J. O'Connor, chair of Com Ed, declared that the capacity cap and some other proposals were "hostile" to the utility and were designed to "punish utility stockholders" (Rosenheim 1985).

FIG. 4.3 **Edison Chairman and Chief Executive Officer James J. O'Connor**

The Illinois legislature completed its revision of the Public Utilities Act in July. The amendments were to become effective on January 1, 1986. The legislature adopted most of the proposed changes but, heeding Edison, voted narrowly to delete the "excess capacity" provision. The General Assembly chose instead to modify the Act's definition of "used and useful" and to authorize the Commission to make excess capacity determinations on a case-by-case basis. The Assembly declared

The New Rhetoric of Least-Cost Utility Planning

The Public Utilities Act of 1985 encouraged energy conservation and least-cost electric utility planning. The concept of "least-cost utility planning" emerges from the basic observation that electric power generation, transmission, and distribution wastes two-thirds of the primary energy used to produce a usable kilowatt hour of electric power.[32] Though differing on details, the advocates of least-cost planning agree that its fundamental purpose is to deliver *energy services* (not electricity) to consumers at the least total cost to society. Claiming that people consume not so much electricity as the services that electricity helps facilitate, these advocates define energy services as the *end-use activities* and conditions that energy helps the consumer do or experience. Thus mobility, comfort, light, cleanliness, and food preservation and preparation are the kinds of end uses that householders are likely to demand. These end uses are provided by a combination of energy and end-use technologies (e.g., air conditioners and refrigerators).

Advocates of least-cost planning argue that the future demand for electric power can best be forecasted through *end-use engineering models* that explicitly account for consumer adoption and use of end-use technologies. They also argue that utilities should compare supply- and demand-side investments on an equal basis, choosing a resource mix that minimizes total costs. In their view, *supply and demand planning should be integrated*: the cost of supplying an additional kilowatt hour of electric energy should be compared with the cost of conserving or avoiding a kilowatt hour, with the utility investing in conservation and generation options in optimum economic order.

Thus advocates of least-cost planning reject the radical separation of production and consumption that underlies the institution of regulated natural monopoly. They seek to blur the boundaries between the two.

that "a generation or production facility is used and useful only if, and only to the extent that, it is necessary to meet customer demand or economically beneficial in meeting such demand" (Illinois Supreme Court 1991, 1051).

Completion of the Byron 1 plant provided the first test of rate making under the new law. In November 1984, Com Ed applied for a 12.2 percent rate increase to yield $544.6 million. For the most part, this increase reflected the anticipated costs of including Byron 1 in the rate base. Construction of that unit was actually completed in August of the following year, six years behind schedule and $3.7 billion above the originally projected cost. The extraordinary delays and cost overruns

FIG. 4.4 **The Completed Byron Nuclear Power Plant**

forced the ICC to ask whether any part of the cost of the unit should be excluded from the rate base because of imprudence or managerial inefficiency by Edison. A key element in the Commission's deliberation was the Arthur D. Little audit of Byron 1 construction costs. Completed in March 1985, the audit found substantial cost overruns at Byron 1 and 2, attributing them to continual changes in NRC safety and construction requirements and to deficiencies in Com Ed's monitoring of construction. BPI and the City of Chicago argued that the audit was not conducted in accord with "generally accepted auditing standards," as required by the recently amended Public Utilities Act. Upon reviewing the audit and related testimony, however, the Commission rejected that argument. On a 4–3 vote, it decided to attribute half the costs of the 1984 licensing delay to Edison and to reduce the cost of Byron 1 for rate base purposes by $101.5 million. The fair value of Edison's rate base became $8,884.9 million, and the fair and reasonable rate of return became 12.63 percent. On this basis, the ICC granted Edison a rate increase totaling $494.8 million, or 11.0 percent, to be phased in over four years. The Commission also instructed Edison to prepare proposals for phasing Byron 2 and Braidwood into its rate base (ICC 1985).

Consumer groups were deeply upset by the Commission's rulings. "This is one of the most outrageous decisions ever handed down by this commission," steamed Susan Stewart of CUB. "It represents a total disregard for the facts and evidence of the case." She declared that "we have no choice but to appeal this decision in court" (Schneidman 1985a). And when the two new members of the Commission, Meg Bushnell and Susan Stone, vigorously dissented from the majority decision, they focused their displeasure on the audit's failure to separate reasonable from unreasonable costs (ICC 1985b).

The ICC paid close attention to the words *prudence* and *reasonable* in its Byron 1 decision. According to the 1985 Public Utilities Act, *prudency* concerns whether a utility's initial decision to construct a new facility and any subsequent reevaluations of that decision were sound. The Commission is to determine prudence on the basis of evidence and information which was known or should have been known at the time of certification, initiation of construction, and each subsequent evaluation. *Reasonable,* on the other hand, "means that a utility's decisions, construction, and supervision of construction . . . resulted in efficient, economical, and timely construction. In determining the reasonableness of plant costs, the Commission shall consider the knowledge and circumstances prevailing at the time of each relevant utility decision or action." A decision to build a unit might be prudent yet still yield costs that are not entirely reasonable, and only costs that were both prudently incurred and reasonable could be included in the rate base. (See ICC 1986a.)

Once approved, the Byron 1 rate increase became part of the 1986 gubernatorial campaign. In October 1985, Attorney General Neil Hartigan, a Democrat, said that he would ask the ICC to reconsider its decision. If it refused, he threatened to file suit to block the increase. "This is Jim Thompson's commerce commission," he said. "Phil O'Connor [then chair of the ICC] was Thompson's campaign manager. Anyone who doesn't think that this is the governor's rate increase is kidding themselves." Adlai Stevenson, Thompson's prospective Democratic rival for governor, condemned "the governor's commerce commission" for approving the increase and Hartigan for not fighting it when it was before the Commission. Even Republican Governor Thompson said he was "not satisfied" with the increase and maintained that he had criticized it before Hartigan. Each blamed the other for the increase (Wingert and Gratteau 1985).

On October 31, the ICC refused to reconsider its decision or to halt the increase. A group of intervenors—including the Attorney General, the Governor, the Cook County State's Attorney (Richard M. Daley), the City of Chicago, BPI, and six others—appealed the ICC's decision to the Cook County Circuit Court. Douglas Cassel, Jr., of BPI said of

FIG. 4.5 **The Illinois Commerce Commission in Late 1985: Standing from Left, Commissioners Barrett, Levin, O'Conner, Kretschemer, and Manshio; Seated from Left, Stone and Bushnell**

the Commission's decision, "It is analogous to being on a train. We have passed the LaSalle stations, and we are pulling out of the Byron stations, looking forward to hitting the Braidwood stations, while the City of Chicago gazes wistfully out the parlor-car window, toying with the idea of getting off the train completely" (Schneidman 1985b).

Just two months after the Byron 1 decision, Philip O'Connor and Stanley Levin resigned from the Commission. O'Connor had announced his intention to resign in early October, saying that "there are no major issues to come before the commerce commission in the next two years," and that he "would take no job that would give the appearance of conflict of interest" (Schneidman and Egler 1985). Levin had been pressured to resign for actually residing in Missouri (Schneidman 1985c). Apparently responding to the public outcry against the rate increase, Governor Thompson created a more consumer-oriented commission by appointing two new commissioners (Paul Foran and Richard Romero) and naming Meg Bushnell as the new chair.

In December 1985, during ICC hearings concerning Braidwood, Com Ed stoked the fire created by the Byron 1 rate increase by announcing that its last three nuclear units would cost $1.2 billion more and take eight to nine months longer to complete than last estimated. The cost of Byron would increase by $281 million to $4.65 billion and of Braidwood by $940 million to $5.05 billion. Two months later, the company responded to the ICC's October request for nontraditional alternatives. Under traditional rate making, Com Ed would ask for an 18.6 percent increase in the first year ($1.04 billion), followed by 9.3 percent ($630 million) and then 3.8 percent ($290 million). Saying that it wanted to minimize "short term effects and avoid so-called rate shock" (Schneidman, 1986a), the company proposed three nontraditional ways to phase in rate increases for these last three nuclear units:

- 4.8 percent per year for eleven years.
- 6.4 percent in the first year, followed by increases 0.5 percent lower than that for the next ten years, and an 11.4 percent reduction at the end of eleven years.
- 6.8 percent per year for seven years.

Rates would stabilize after that, according to Edison. Howard Learner of CUB, however, called the options "a sham." "[W]hat it boils down to is that Edison is asking: 'Do you want to take your poison in one big gulp, or do you want to take more poison over a longer period of time and in smaller doses?'" (Schneidman 1986a). Mayor Washington declared that a traditional 35 percent increase would "deliver a crippling blow to the economic health of Chicago" (Schneidman 1986b).

Commonwealth Edison had been shocked by Black January. In early May, it received another, potentially worse, shock. News leaked of the massive nuclear accident on April 26 at Chernobyl in the USSR. Like others in the nuclear power industry, Edison officials argued that design differences prevented any Chernobyl-type accident at an American nuclear plant. Still, public confidence in the nuclear power industry was deeply shaken. In mid-June, Daniel Ford, author of a major book about Three Mile Island, urged Com Ed to shut down its aging Zion nuclear plant and to halt construction on Braidwood. Emphasizing the risks of a meltdown near Chicago, he warned that "the Russians aren't the only ones playing Russian roulette with nuclear power. Commonwealth Edison is also doing it" (Bukro 1986).

In April 1986, CUB proposed additional legislative and regulatory reforms. These would exclude the costs of building unneeded plants and establish a trust fund for future dismantling of nuclear plants. The legislation also would create a consumer intervention fund to help public interest groups challenge requested rate increases, and it would

How Many Deaths from Chernobyl?

Chernobyl ended the debate about whether a "worst case" nuclear accident could happen. Thirty-one people died immediately after the accident, and another 206 suffered acute radiation sickness. Roughly 115,000 people were evacuated; 24,000 had received "very substantial doses" of radiation. Estimates of long-term health damage vary widely. Some scientists estimate a few thousand cancer cases worldwide over the next fifty years; others, more than a million. Anspaugh, Catlin, and Goldman (1988) estimate that as many as 17,400 radiation-induced fatalities might result from the accident, a total "undetectable" in comparison with the 513 million "natural or spontaneous" cancer fatalities expected over the next fifty years. They also estimate the total economic cost of the accident to be about $15 billion. In contrast Dr. Vladimir Chernousenko, the scientific supervisor of the emergency team sent into Chernobyl a few days after the meltdown, believes that the accident killed at least five thousand "cleanup" personnel and injured (or will injure) 35 million people (Matthiessen 1991).

Proponents of nuclear power in America argue that Chernobyl's graphite reactor had basic design flaws, including the lack of a containment shell, and that a similar accident was highly improbable at an American facility. Opponents argue that the accident demonstrated that it is not possible to operate a "safe" nuclear reactor (e.g., Haynes and Bojcun 1988). Dr. Robert P. Gale, who performed bone marrow transplants on some accident victims reportedly said: "It's one thing to talk about deserted zones and another to drive into a 3,000 square-kilometer area just completely devoid of human life, to fly over an entire forest which is brown because the trees are radioactive. It has a tremendous emotional impact. It's unavoidable when you see these things that you understand the potential dangers of radiation" (Broad 1987).

require utilities to indicate clearly whether their advertisements were paid for by shareholders or ratepayers.

Shortly thereafter, Judge Richard L. Curry of the Cook County Circuit Court overturned the ICC's decision on Byron 1. He ruled that the Arthur D. Little audit had not been conducted under "generally accepted auditing standards" and that the Commission had failed to require Edison to prove that the construction costs had been "reasonable" as required by the 1985 Public Utilities Act. He also ordered the ICC to exclude from Edison's rate base all the costs of the 1984 licensing delay, to exclude some portion of the costs of the plant common to Byron 1 and 2 from the costs of Byron 1, to roll back the rate increase approved in 1985, and to set revised rates within thirty days (cited in Illinois Supreme Court 1987). In mid-May, Curry refused to delay his order or to drop the thirty-day deadline. He did allow Com Ed's increase to remain in effect while the company appealed his order, but required it to file a record of the money collected so that refunds could be made. Com Ed appealed Curry's decision directly to the Illinois Supreme Court.

By now, the rate increase for Byron 1 was in trouble. But what of Bryon 2 and Braidwood? Their construction continued—despite the fact that, since late 1982, the ICC had been investigating the economics of energy conservation as an alternative to Braidwood. In late 1982, BPI asked the Commission to assess energy conservation alternatives for Edison. A little over a year later, it asked the Commission to consider canceling Byron 2 and both the Braidwood units. The ICC deliberated until mid–1986. And while the ICC deliberated, Edison continued to build.

BPI and other consumer intervenors were claiming that Braidwood represented unneeded capacity and that it would be much cheaper for Edison to invest in energy-conservation measures. Cancellation of the Braidwood plant would, they argued, save $3.35 billion in direct costs—and probably more, indirectly. Amory Lovins (the nation's best known advocate for energy conservation) chimed in with September 1985 testimony for BPI that Com Ed could save ratepayers $3-7 billion by canceling Braidwood and that Edison's customers would be trapped in a death spiral of ever-increasing rates if the plant was not canceled. Edison replied that Braidwood was a reasonable and necessary addition to its generating capacity. It argued that completion of Braidwood would result in approximately $2.7 billion in benefits (ICC 1986b).

These stunningly different estimates derived almost entirely from differences in assumptions about five variables: Braidwood capital costs, load-growth forecasts, Braidwood operation and maintenance costs, Braidwood capital additions costs, and Edison's ability to earn a return on capital invested in a plant that might not prove used and useful. In light of the conflicting estimates, the Commission decided in July 1986 to (1) cap Braidwood construction costs at $5.05 million, (2) consider later whether any portion of those costs had been imprudently incurred, (3) order Edison to show cause within thirty days why its certificate of convenience and necessity for Braidwood 2 should not be withdrawn or altered, and (4) instruct ICC staff to begin an immediate investigation of alternatives to traditional methods for treating new plant costs (ICC 1986b).

In arriving at this decision, the Commission expressed alarm about the dramatic cost increases and delays associated with Edison's nuclear units. "Under these circumstances," the Commission declared, "neither equity nor the law requires this Commission to be bound by Orders in dockets which were based on evidence presented by the Company itself which history has shown to be notably inaccurate" (ICC 1986b, 29). The Commission also expressed some frustration at being "severely constrained by the evidence as it has been presented by all parties in the proceeding" and being authorized "only on the narrow issues of whether both Braidwood units should be cancelled or whether the Certificate . . . previously issued to Edison to complete Braidwood should be withdrawn" (29–30). Lastly, the Commission noted the uncertainty that "shrouds" much of the evidence and testimony concerning several crucial issues. Noting that "all parties to this proceeding have cautioned the Commission against ascribing a false and nonexistent precision to some of the major forecasts and projections, including each party's own projections, which affect the decision to cancel or complete," the Commission "declines to attribute to this evidence a reliability which the parties themselves are unwilling to acknowledge" (30). Focusing on the parties' mathematical models, the Commission stated:

> Although Edison and the Petitioners have both provided mathematical models, each supporting their respective positions regarding future load growth, the Commission is not convinced that either model is more accurate than the other. . . . We believe that forecasts may provide useful indications of what may occur; however, they are only one piece of evidence to be considered when analyzing the testimony in the present docket. Positing formal mathematical models is useful in understanding why a party makes the estimates it makes. However, exposing one's reasoning does not validate one's conclusions, it merely helps explain how the conclusion was reached. (ICC 1986b, 36)

Here the Commission comments about the irony of having to rely on forecasts that all parties acknowledge are unreliable. At a deeper level, I sense the Commission wondering how the collective story about planning and rate making in Illinois will unfold when it is being shaped by authors with conflicting points of view and conflicting plots. Neither Edison nor consumer groups could control or even understand fully the variables that determine future demand.

Shortly after this decison, the ICC ordered an audit of Braidwood 1's construction costs, as required by the new Public Utilities Act. By a 4-3 vote, the ICC awarded the contract to O'Brien, Kreitzberg and Associates of San Francisco. At that point, the construction of Braidwood 1 was almost finished. Braidwood 2 was approximately 60 percent complete and scheduled to begin operation within two years. But would the units be needed? Sales had grown at an average annual rate of 1.0 percent per year from 1974 through 1985 (and peak load by 0.6 percent), far below the rates originally forecast back in the early seventies.

Whether to cancel the Braidwood units (or at least Braidwood 2) was an issue inseparable from the adequacy of Edison's methods for forecasting sales and peak loads. Both, in turn, were tied intimately to the adequacy of Edison's program for conserving energy. In September 1986, the ICC concluded two separate investigations of Edison's forecasting methodology and energy conservation.

Modernist planning seeks to control the future, bending it to the will of the present. Like other electric utilities, Edison used to forecast future demand simply by *extrapolating from past trends* (much as I did in chapter 3). The 1973 oil embargo and subsequent decline in demand induced them to develop alternative approaches. Edison and many other utilities began using *econometric models.* Such models attempt to explain or predict the movement of one variable (e.g., peak load) in terms of the prices of electricity and competing fuels, consumer income, population, and other demand-influencing variables. One crucial assumption of econometric modeling is that historical relationships between the independent variables and the demand for electricity will continue into the future. It assumes, in other words, that the plot of the modernist story will continue. Figure 4.6 below suggests that such an assumption was unwarranted in 1986.

In the late 1970s, consumer advocates and a few other utilities began developing a modeling approach that implies a new planning story. Referred to as *end-use models,* and particularly applicable to the residential and commercial sectors, these models relate the demand for electricity to the current stock of energy-using capital, then forecast changes in the demand for power by forecasting changes in the number of devices over time. Note, though, that this modeling approach presumes an ability to foresee how legislative and regulatory action will affect the incentives to use energy-conserving devices. Edison has strongly resisted consumer groups' arguments that they rely more heavily on end use models.[33]

The docket on load forecasting had actually been opened in 1980 as an examination of the company's future (post-Braidwood) construction program. In the face of a declining rate of growth in demand, the company decided not to plan any additional construction beyond Braid-

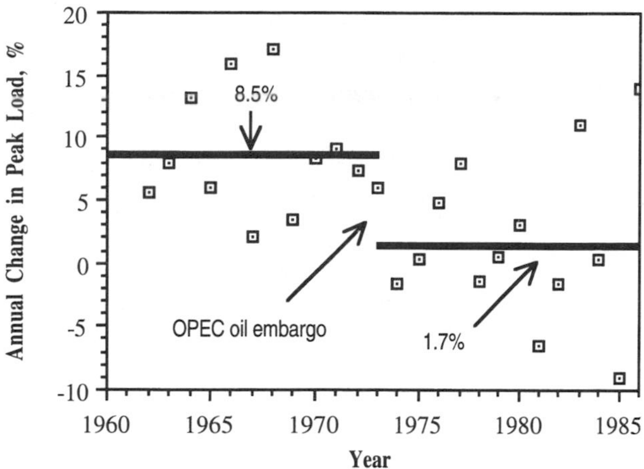

FIG. 4.6 **Annual Growth in Peak Load Drops Precipitously after 1972**

wood. This induced the ICC to focus its investigation on Edison's methodology for forecasting load growth. The inquiry revealed that as of 1986, Com Ed had relied upon two internal groups to develop its official forecasts of load growth. Its Statistical Research Staff developed the models that the company used to forecast growth in peak load and generated the data that were input to those models. Its Load Estimates Committee, comprised of twelve to fourteen of the company's senior managers, evaluated the statistical staff's results and adopted the company's official forecast by consensus. Consumer groups recommended that Edison be required to provide complete documentation of its forecasts, begin developing end-use load forecasting capabilities, and begin implementing cost-effective alternatives for meeting any future growth in load. Edison asked the Commission to find that the company's load-forecasting methodology "is reasonable and that the Company is acting reasonably to update its methodology to keep pace with changing conditions" (ICC 1986c, 465).

In the end, the Commission observed that it "does not know how and if . . . the load forecasting process and the generating expansion process, are meshed at the Company" (467). Moreover it rejected Edison's request to rule the company's methodology reasonable. The Commission concluded that "supply and demand side alternatives to capacity expansion should be considered when preparing plans to meet projected increases in future peak load" (477). It also said that the company's current econometric models should be improved in several specific ways and that the company should develop an end-use model for the residential sector.

The ICC concluded its three-and-a-half year investigation of Edison's energy conservation programs in late 1986. Edison's vice president of load management, conservation, and marketing testified that conservation was not a cost-beneficial option for the company in the near future. This was largely because the company had low fuel costs and ample generating capacity. Consumer groups, on the other hand, argued that consumers and society as a whole could achieve substantial savings by having Edison pursue an aggressive conservation program. As part of their testimony, consumer groups had invited Amory Lovins to present a highly detailed description of new energy conservation technologies. Arguing that immense savings could be obtained, Lovins encouraged the Commission to adopt conservation as an alternative to further supply-side construction.

Lovins's testimony challenged the mainstream view that future growth in the ecomony depended on growth in electric power supply. It was true, he said, the electricity demand and Gross National Product (GNP) had been highly correlated with one another for many decades. However, he emphasized that the ratio of electricity demand to GNP (which measured the national economy's aggregate electrical intensity) had been declining since 1977. True, some of that decline could be attributed to recent increases in the real price of electricity. But Lovins argued in response that further declines did not depend on continued real price increases since "efficiency improvements . . . generally cost far less even than *present* electricity prices" (p. 39). Furthermore, though it was true that electricity had been gaining an increasingly large share of the energy market, that did not necessarily mean that the demand for electricity would increase in the future since "electricity could well have an increasing share of a dwindling market" (p. 41). Lovins further argued that the electricity content of a dollar of real GNP was likely to fall even faster that it had through the early 1980s:

> [H]ow soon and how steep that fall is will depend on public and private policy. . . . For it is a basic premise of my testimony that *demand for electricity is not fate but choice. Demand is not a predetermined outcome to be prophesied by reading the entrails of forecasters, but rather a variable to be influenced in accordance with policy goals.* Demand is therefore not to be forecast so much as managed. (1985, 44)

Though impressed by much of Lovins's testimony, the Commission concluded that "many of these technologies are either not available to the general public, due to market or financial reasons, or are not necessarily adaptable to the weather conditions or customer demographics of Edison's territory" (ICC 1986d, 143). The Commission decided that

TABLE 4.2 **Eleven Years of Rate Increases**[a]

Date of ICC Decision Granting Increase	Rate Increase Sought/approved ($ million)	Rate Increase Sought/approved (%)
Apr 10, 1974	154.0/134.7	12.2/10.7
Aug 27, 1975	241.0/207.0	15.6/12.7
Oct 12, 1977	263.0/151.0	14.5/7.6
Dec 13, 1978	125.0/74.9	5.6/2.7
Feb 6, 1980	452.0/389.6	18.3/14.4
Jul 1, 1981	682.0/503.6	19.7/14.5
Dec 1, 1982	805.0/660.7	19.4/17.1
Jul 17, 1984	462.2/282.5	10.7/6.6
Oct 24, 1985	544.6/494.8[b]	12.2/11.0[b]

[a] *Sources:* ICC, 1974, 1975, 1977, 1978, 1980b, 1981, 1982b, 1984, 1985.
[b] The increase was phased in over four years. Rates would increase by 9.2% in 1985, rise to 12.6% in 1987, then decline to 11.0% in 1989.

> it would not be prudent to institute full-scale conservation programs for one utility based on estimated energy savings and costs from other jurisdictions, without any evidence that the program would not meet unanticipated market, financial or psychological barriers in Illinois that could have been avoided or eliminated based on data gathered in lower cost, smaller pilot programs. . . . [Nonetheless,] this Commission is very concerned that these pilot programs be completed and conservation, as a regular part of Edison's operations, be implemented as quickly and economically as possible. (ICC 1986d, 143–44)

Guided by this perspective, the ICC ordered Edison (by a 4–3 vote) to begin a rather modest nine-point energy conservation program. Two commissioners unsuccessfully argued that "the issue is whether government should induce the consumer to a particular course of action or whether the consumer should exercise his/her freedom of choice and be responsible for that choice" and that energy conservation "negates the freedom of choice now available in satisfying energy needs" (ICC 1986d, 178).

In mid-September, Chicago, CUB, BPI, and others filed a motion with the ICC to cancel Braidwood 2 immediately. They argued that the ICC should not wait to hold hearings in the fall. But the ICC voted 7–0 to delay a decision on that unit for another five months. Meanwhile, Edison continued to build its last three nuclear units. Both major party candidates for governor, Thompson and Stevenson, said they favored cancellation of Braidwood 2. Thompson also said he favored banning the construction of any more nuclear units in Illinois, urging

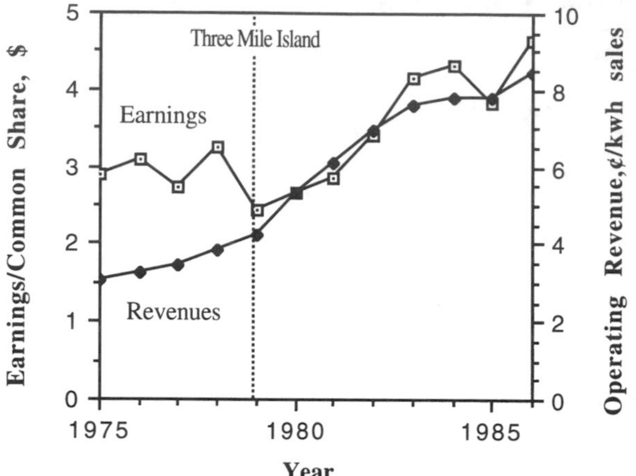

FIG. 4.7 **Edison's "Price" and Earnings Rise in Early 1980s**

Source: Moody's (1977, 1984, 1988).

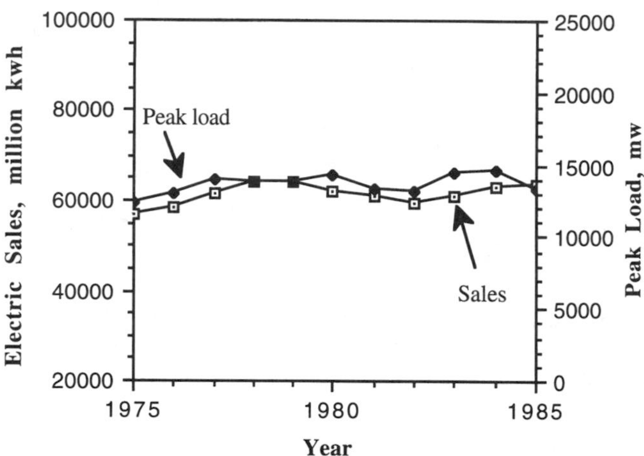

FIG. 4.8 **Sales and Peak Load Growth Stagnate in Early 1980s**

Source: Moody's (1977, 1984, 1988).

instead that Edison and other electric utilities exploit the state's ample coal reserves.

By late 1986, it was not at all clear whether Commonwealth Edison would be able to obtain a rate increase for its last three nuclear units. It had almost finished the ambitious construction program started in the early seventies, and it had been able to obtain most of the rate increases requested in the late seventies and early eighties (see table 4.2).

Fueled by these rate increases, the company's operating revenues per kilowatt-hour and earnings per share had improved markedly (see figure 4.7). Yet there was a major downside. The public appeared to have been angered by the long series of costly rate increases. The major elected officials had responded to that anger by appealing the Byron 1 rate decision and by publicly opposing completion of Braidwood 2. The current Illinois Commerce Commission seemed to favor the consumer point of view more than its predecessors had. And the company's sales and peak load had stagnated (see figure 4.8). The modernist approach to rate-making had broken down.

On December 20, 1986, Governor Thompson, Attorney General Hartigan, State's Attorney Daley, and Edison management announced that several weeks of private negotiations had enabled them to hammer out an innovative rate proposal. The proposal was, according to the Governor, "one of the best deals for Illinois consumers that I have seen in my 27 years of public service and 10 years as governor" (Eissman 1986a).

Was this the "best" or even a good deal for Illinois consumers? How would the question be answered in the context of the modernist story of electric power planning and rate making? Those questions preoccupied Edison, the ICC, and the major consumer groups throughout the first half of 1987.

Interlude

The Illinois Commerce Commission's Offices in Springfield

It's a cold, dreary, blustery morning. Certainly no surprise for the first day of spring in Illinois. I walk past the State Capitol, then turn toward the Illinois Commerce Commission's building three blocks to the east. There it is: a nine-story brown brick building, with little external embellishment. Stepping inside, I recall the days when I worked for a governmental agency in Louisville, Kentucky. The recollection discomforts me. I believe in democratic governance, in the principle of a community of people governing themselves. But here, as in Louisville, I sense something different. The building feels like a secure place in which to make a monotonous but necessary living. "Not a place for ambitious, energetic, congenial people," I think. But who am I to judge? Focus on my own work. There to review the Commission's files for Edison rate cases, I am dismayed to learn that I have to pay twenty-five cents per page to copy material. "How's a person supposed to learn about a case if he has to pay that price?" I wonder. After reading through seemingly endless pages of microfiche documents, I need to move my aching limbs. Downstairs I go, to the Commission's main hearing room. The room is roughly 30 feet long and 30 feet wide. Its ceiling is twelve to fourteen feet high, banked with fifteen fluorescent lights that are separated by broad wooden decorative beams. No windows break the monotony of the cream-colored walls. Except in the back, and even those windows are curtained. Metal chairs with plastic upholstery sit erectly on a dark brown floor. The room is quiet until the commissioners, staff, counsels, and others enter. I notice no distinct aromas. It is a clean, well-lighted place, just as you would expect for a seventy-year-old regulatory commission.

5

Once the plan is fully understood, it will be evident that it's the best deal for consumers.

— A COMMONWEALTH EDISON OFFICIAL

The Best Deal for Illinois Consumers? Assessing Commonwealth Edison's "Negotiated Settlement"

Edison had prepared a very ambitious nuclear expansion plan in the early 1970s. Unfortunately, that plan became highly controversial and difficult to carry out completely or on time. Edison's nuclear units took much longer to build, and cost much more, than planned. Furthermore, demand growth had tailed off dramatically and it was not at all clear that the plants would be needed before the mid- to late 1990s, if then.

Edison and consumer groups configured those delays and overruns in terms of two different stories. Defining themselves as expert managers of an extraordinarily complex enterprise, Edison officials insisted that those delays and cost overruns were caused by factors beyond their control, especially government regulation driven by politics and other irrational forces. Conceptualizing the industry in the modernist terms of public utility economics, Edison officials argued that they had initiated construction of the Byron and Braidwood units in order to meet projected growth in demand, provide adequate reserve capacity, and maximize long-run economic efficiency. In their view, the company was entitled to earn revenues that reflected the costs of building, financing, and operating those plants. In early 1986, they had announced that the company would probably need another 32 percent rate increase to cover the roughly $7 billion cost of its last three nuclear units. Edison officials knew that such a massive increase could produce a "rate shock" and lead to even more litigation over its rates and construction program, so—in mid-1986—they began to explore alternatives to traditional rate base treatment of the last three units.

Consumer groups told a different story. At the core of the coalition that opposed Edison's requests for rate increases was a set of community, consumer, antinuclear, environmental, and labor groups, including CUB, which represented residential ratepayers in utility regulation issues; BPI, a public-interest law firm that represented low-income consumers and environmental concerns; and numerous community organizations that sought to mobilize and empower low-income communities. These groups generally saw Edison as a grossly mismanaged relic of the past that—in a shortsighted pursuit of profit—had drastically overbuilt its power-generating capacity, made northern Illinois one of the highest-cost electric power regions in the country, driven businesses and jobs away, and greatly increased the risk of nuclear accidents.

Claiming that Edison was building far more generating capacity than its customers would need anytime soon, consumer groups petitioned the ICC to order cancellation of the Braidwood plant and to support the groups' position on three related rate cases, and they appealed Edison's 11 percent rate hike for Byron 1 to the Illinois courts.[34] They also successfully pressed for adoption of a new Public Utilities Act in 1985.

The new Public Utilities Act formed a critical part of both stories. Through it, the consumer groups and the General Assembly had changed the rules of the game under which Edison was regulated. The new law strengthened the ICC's regulatory powers, created a new "least-cost" electric utility planning process, facilitated investment in energy conservation measures, gave citizens standing in the court to sue the Commission for any breaches of its duty, and required the ICC to audit the construction costs of new plants before those costs could be included in the rate base (Platt 1987). The audit provisions were crucial; they threatened to cause Edison to lose millions that the utility did not regard as at risk when it began its nuclear construction program.

These contending stories also had electoral implications for politically ambitious officials, especially Governor James Thompson. Up for reelection in November 1986 and aware of intense consumer hostility to Edison's past and impending rate increases, Thompson took several steps to improve his standing with voters. He too appealed the Byron 1 decision (along with Attorney General Neil Hartigan and Cook County State's Attorney Richard Daley). He then created a more consumer-oriented ICC by appointing two new commissioners and by installing Margaret "Meg" Bushnell, a vocal consumer advocate who lived near the Byron plant, as the new chair in December 1985.

Then, shortly before the election, Thompson declared that the Braidwood plant should be canceled and that no new nuclear units should be built in Illinois. These moves apparently succeeded, for Illinois voters reelected Thompson. Even so, it was not clear how he would deal with the last three nuclear units and the stories that focused on them in late 1986.

The "Best Deal" Press Conference

"This is one of the best deals for Illinois consumers that I have seen in my 27 years of public service and 10 years as governor," Governor Thompson told a crowded press conference on December 20, 1986. Joined by Attorney General Hartigan, State's Attorney Daley (both Democrats), and Edison officials, the Governor was referring to an innovative electric power rate proposal that the three politicians and the utility had reportedly hammered out over several weeks of private negotiations (see fig. 5.1).

The "best deal" was complex and multifaceted. On the one hand it would allow Edison to place its last three nuclear units into a new generating subsidiary (GENCO). Com Ed would pay GENCO $660

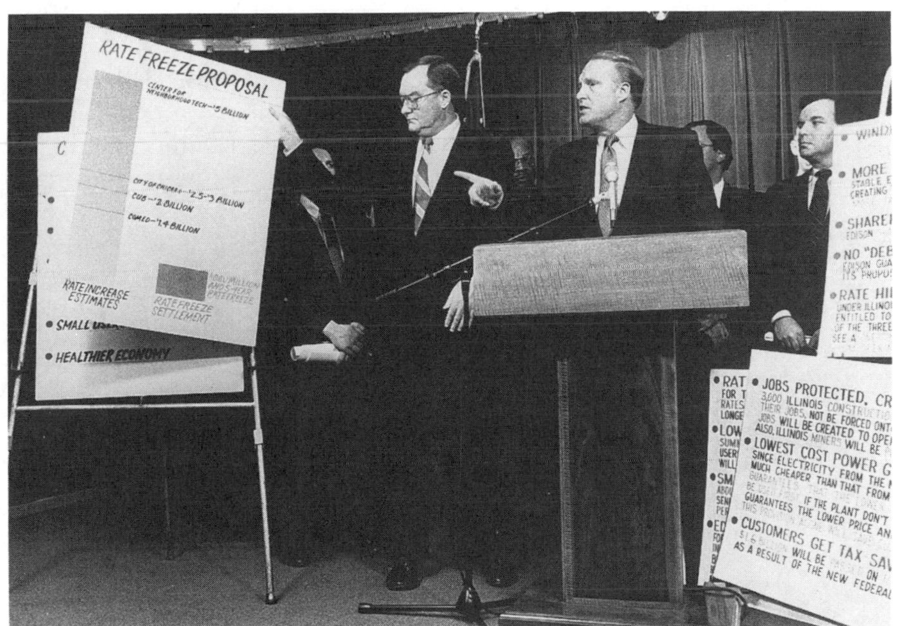

FIG. 5.1 **Governor Thompson, Left, Attorney General Hartigan and State's Attorney Daley, Right, Discuss Edison's Innovative ("Best Deal") Rate Proposal**

million per year for the next five years, in return for which the subsidiary would sell its power exclusively to Edison. At the end of five years, the annual payment would cease and the subsidiary would be free to sell power on the open market. To recoup the $3.3 billion involved in those five payments, Com Ed would be allowed to raise its rates by $660 million, or 12.9 percent. Rates would then be frozen at those levels for five years (barring any unforeseen events or "acts of God"). By transferring the three new reactors to the subsidiary, furthermore, Edison would shift regulatory responsibility for those units from the ICC to the Federal Energy Regulatory Commission (FERC); the ICC would automatically pass along the costs of any subsequent FERC-approved rate increases to ratepayers. The transfer would also exempt the last three plants from the construction cost audits that would otherwise be required by the 1985 Public Utilities Act. Consumers would benefit substantially from this deal, Com Ed and the three officials agreed; rates would be frozen at the new levels for the next five years, thereby stabilizing rates and making them more predictable. In return, governmental agencies and consumer groups would drop the Byron 1 Supreme Court appeal, the Braidwood cancellation case, and their other cases pending before the ICC.

The proposal caught consumer groups by surprise. Though the deal would require them to drop their ICC cases and court appeal, none of them had been asked to take part in the "negotiations." Believing that the three nuclear units were not needed, contending that legal loopholes might free Edison from sticking to the agreement, arguing that Com Ed would try to recover any revenues lost during the rate freeze by increasing rates after five years, they attacked the proposal and the people behind it. Speaking to reporters after the press conference, Howard Learner, a young activist attorney who had spearheaded the creation of CUB in 1983 and who was also a counsel for BPI, said, "It's a mine field for consumers, it's smoke and mirrors, the worst deal I've ever seen" (Eissman 1986a).

FIG. 5.2 **BPI's Howard Learner**

Unmoved by these complaints and threats, Com Ed officials responded that "There is no room for negotiation . . . this is the result of months of negotiations and this is our maximum bend . . . [and that]

Not getting the Supreme Court case dropped is a deal-breaker" (Eissman 1986a). Edison was, however, not in a position to dismiss the consumer groups unilaterally. It needed the ICC's approval. And the chair of the ICC, Meg Bushnell, wanted Com Ed to negotiate with the consumer groups. "They've got to give on their position," she said. "They can have the support of Daley, Hartigan and whoever else it has, but it needs the support of these [consumer groups]. These groups have legal remedies and no one, no matter who they are, can make them drop their suit" (Eissman 1986b).

Despite Bushnell's initial reaction, Edison remained firm and unyielding: it would rather withdraw its proposal than negotiate with consumer groups. "We've given our absolute best on this agreement now," said one of its vice presidents, "and we just can't change it." Spokesmen for Thompson, Hartigan, and Daley agreed. "We have determined this is the best, most comprehensive deal for consumers and so we're for it," said a spokesman for Daley; "if it could be changed for the better, it wouldn't [already] be the best" (Eissman 1986c).

Com Ed's stance was "silly," Bushnell said; "it's arrogant not to talk to people . . . not to negotiate with them" (Eissman 1986c). Hearing this, Com Ed quickly reversed itself—thus beginning a six-month-long process of tortured backpedaling from its initial position—and said that it was willing to negotiate with consumer groups. "Everything in life is negotiable," said a vice president. "We will talk to any of the groups, listen to their suggestions, and if they improve the agreement, we will incorporate them" (Eissman 1986d). "Once the plan is fully understood, it will be evident that it's the best deal for consumers" (Eissman 1986e).

Private Negotiations

Was this the "best deal" for consumers? One way to find out was for Edison and its allies to follow Meg Bushnell's advice and negotiate with consumer groups directly. Attorney General Hartigan brought the conflicting parties together in late December to analyze, discuss, and negotiate a revised proposal; and—unlike Governor Thompson and State's Attorney Daley—he sat in on a number of the negotiating sessions as they proceeded over the next forty-seven days (People of the State of Illinois 1987; Tanzman 1990).

At first Edison and the consumer groups were far apart (see table 5.1). Edison wanted the terms specified in its original "best deal," but consumer groups insisted that consumers should not have to pay for the three new plants, abandon the suit about Byron 1, drop their three ICC cases, or accept any escape clauses about "acts of God," regu-

TABLE 5.1 **Initial Negotiating Positions**

Commonwealth Edison	Consumer Groups
Create GENCO regulated by FERC	ICC regulation; no GENCO
Drop Supreme Court case	Not drop the Supreme Court case
\$660 M rate increase; freeze rates for 5 years (except under certain conditions)	No rate increase
Consumers drop 3 ICC cases	Not drop the 3 ICC cases
No cost audits on 3 newest nuclear units	Audit costs of 3 newest units

latory or legislative changes, or economic disruption. They also charged that, by proposing to create a generating subsidiary, Edison was trying to shift regulatory oversight to an agency (FERC) that would be more favorable to the utility.

In mid-January, just a few days before Com Ed intended to present its formal proposal to the ICC, consumer representatives publicly announced their counter proposal to Edison's original "best deal." The consumers' counter would have Edison drop its plans for a rate increase followed by a five-year freeze. In return CUB and BPI would withdraw their Supreme Court suit and ICC cases. Upset at seeing the counterproposal made public, Edison officials, Hartigan, Thompson, and Daley declared it "unworkable" (Biddle and Johnson 1987). These first responses notwithstanding, the contending parties had by the end of January come to agreement on most of the major points of initial disagreement. Only one remained: Edison clung tenaciously to its demand for a \$660 million rate increase while consumer groups pushed for a maximum increase of \$395 million.

After closed-door negotiations that lasted forty-seven days, the negotiations came to an impasse over the size of the rate increase. So, on February 3, Com Ed formally asked the ICC to approve a \$660 million increase, along with other elements of a "negotiated settlement" contained in a "Memorandum of Understanding" signed by the Chairman of the Board of Com Ed, the Governor, the Attorney General, the State's Attorney, and other minor parties. Thompson, Hartigan, and Daley pledged their continued support for the proposal. "We're not fools," said Thompson; "If there was any way we could get more for consumers, I don't think you'd find three men more willing" to continue bargaining. But CUB, BPI, the City of Chicago, and several other consumer groups were unwilling to accept the "settlement" and rejected Thompson's disclaimer. "Edison just refused to budge a dime off its original proposal on price," said Howard Learner; "We were flexible; they were not" (Arndt 1987a).

Evidentiary Hearings

On February 3, seeking to avoid continued uncertainty about Edison's last three nuclear units "without resorting to still further litigation which all agree would only serve to increase the costs to all concerned," Edison and its governmental allies formally asked the ICC to approve their "negotiated settlement" (O'Connor et al. 1987). The Commission assigned the case to two of its hearing examiners. Shortly thereafter the Commission also voted 4–3 to suspend action on the three consumer group cases that Edison wanted dropped as part of its "best deal." A consumer group spokeswoman called the decision "a disaster" (Arndt 1987b).

In early March the hearing examiner proposed a timetable for considering the utility's negotiated settlement. The 1985 Public Utility Act required the ICC to decide the case within eleven months, but the examiner proposed a timetable that would enable the Commission to decide the case within eight. Edison and its allies vigorously objected to this accelerated schedule, arguing that extending the decision past July 1 effectively killed the plan. Two of the three nuclear units would be operational by then, and the company wanted to begin being paid for their construction. They were also worried that the Supreme Court might decide the Byron 1 suit before the ICC acted on the current negotiated settlement. They demanded that the Commission decide within four months.

A few days later the Commissioners voted 6–1 in favor of accelerating the schedule. In doing so they acknowledged that the accelerated schedule might not provide enough time to consider the plan properly, but they also argued that Edison's threat to withdraw its negotiated settlement (and hence leave the Commission without any plan to consider) weighed more heavily for them. Describing Edison's schedule as a "take-it-or-leave-it" proposition, Howard Learner of BPI charged that Edison had "bullied or blackmailed the commission into short-circuiting a full review of the case" (Arndt 1987c).

On March 19, the ICC conducted a public hearing inside the State of Illinois Building in downtown Chicago. Roughly one hundred people spoke at the hearing, with another two hundred people publicly demonstrating against Edison outside. Claiming that "This . . . is the most outrageous railroading job ever perpetrated," Chicago's Mayor Harold Washington told them that greater citizen input was needed, that more alternatives needed to be considered, and that he would go to court to oppose the expedited schedule (Galvan and Unger 1987).

A month later the ICC staff presented its initial analysis of Edison's negotiated settlement. It reported that Edison needed a rate increase of only $180 million, not the $660 million requested, to pay for the electricity generated by the three nuclear units. Edison officials retorted that the staff's report would unfairly keep the utility from recouping any of the construction costs associated with the units (Arndt and Davidson 1987).

From April 21 through May 26, the hearing examiner presided over twelve days of evidentiary hearings in one of the Commission's Chicago hearing chambers. At least thirty witnesses testified, either in direct support of their clients' positions or in rebuttal to other witnesses. The testimony described and evaluated Edison's proposal, but much of it also challenged the degree to which the "settlement" had been "negotiated" and the fairness with which the ICC's proceedings were being conducted. One key claim was that the Governor knew—before announcing his support for Com Ed's rate plan—that the plan was biased in favor of the utility. He had been given a confidential analysis—prepared by a former executive director of the ICC—that indicated that the plan weighted costs in Com Ed's favor and would ultimately enable the utility to recoup all of its costs. A second, and perhaps more damaging, claim was that the utility knew—as a result of detailed technical analyses presented to its board of directors prior to announcement of the "best deal"—that the plan was biased in its favor. A third key claim was that actions by the ICC staff had given Edison an unfair advantage in the hearings. According to this claim, key ICC staff had privately met with Com Ed officials and thereby given the utility five weeks to prepare a rebuttal to a report prepared by the ICC staff. Not offered a similar opportunity to meet with ICC staff, consumer groups had only one week to prepare. The ICC staff member responsible for the Edison case responded that staff would have met with consumer groups if they had only asked. "All they had to do was pick up the telephone and make an appointment," said one of Edison's attorneys (Houston and Davidson 1987). Lastly, the consumer groups charged that the negotiated settlement had not been hammered out in negotiations. "This was Edison's proposal from day one," Howard Learner said; "There were no meaningful negotiations between the government parties and Edison" prior to the date when Thompson, Hartigan, and Daley announced their support (Davidson 1987a).

FIG. 5.3
Citizens Utility Board Consultant Answers Questions During ICC Evidentiary Hearing in May 1987

By early May the consumer advocates had managed to present an unflattering portrayal of both Com Ed and its political allies: Edison had tried to hide a massive rate increase behind the "smoke and mirrors" of a "negotiated settlement," and it had tricked or manipulated Governor Thompson and the others into going along. However, not all of the evidentiary hearings focused on such allegations of political intrigue. Throughout the hearings, expert witnesses attempted to model quantitatively the extent to which the negotiated settlement (as compared to traditional rate making) would allocate fairly to ratepayers and shareholders the future costs, benefits, risks, and uncertainties. Making numerous contestable but unavoidable assumptions, they modeled a wide variety of scenarios which were in turn subjected to detailed sensitivity analyses. Their results varied quite dramatically, largely because of radically differing assumptions about potential disallowances associated with prudency, used and useful, reasonableness, and excess capacity criteria.[35] Two of Edison's allies' witnesses, for example, projected that the negotiated settlement would produce an expected present value savings of $3.24 billion and almost 15 thousand more jobs in the year 2005 (cited in ICC 1987, 506, 507). By varying key assumptions, a consumer group witness projected that ratepayers would save as much as $5.8 billion over fifteen years under traditional rate making (cited in ICC 1987, 519).

ICC staff witnesses also presented expert testimony comparing projections of the present value revenue requirements associated with Edison's proposal and with traditional rate making. The bottom line of their testimony was that the comparison was too close to call and that the decision to accept or reject Edison's proposal should be based on factors not related to quantifiable rate revenue requirements (ICC 1987).

On May 26 the hearing examiners completed their evidentiary hearings. According to their schedule they had roughly two weeks to review the record and prepare a draft order for the Commission. The Commission scheduled oral arguments for June 17 and 18, indicating that it would then accept, reject, or modify the draft order by July 1.

Briefs, Oral Arguments, an Unexpected Event, and the Commission's Decision

On June 1 and 4 the contending parties submitted briefs and reply briefs to the Commission. Edison characterized the proposal as a "settlement" that provided numerous advantages that could not be obtained through "traditional ratemaking" (Com Ed 1987a, 1987b). It was an "unprecedented opportunity," "an historic agreement" that contained "novel and rather startling offers." In the company's view, the parties opposing the settlement presented "biased and one-sided arguments" that were "unremittingly negative." Most importantly, those groups offered the "legally untenable" argument that traditional rate making would result in "draconian rate-slashing" and "huge disallowances of Edison's investment in the Units." The only real standard the ICC should apply, they argued, is whether the "settlement," considered as a whole and supported by the evidence in the record, provided rates that were "just and reasonable." They characterized the ICC's brief as trying to "provide the Commission with an objective and fair evaluation of the Settlement."

BPI (1987) characterized Edison's proposal as a "rate increase" that would violate the Public Utilities Act, that was not the result of "hard-fought negotiations," that was not a "settlement" because not all parties agreed to it, and that would make ratepayers worse off than they would be under traditional rate making. In addition, CUB (1987) characterized it as an effort to "reorganize its way around the law," "a clever scheme designed to allow the utility to recover its investment in the three nuclear units without regulatory interference from the Commission." The Office of Public Counsel (1987) agreed, claiming that "the proposal was cynically packaged in such a way as to obscure its true nature and to stifle opposition" and that Edison insisted on a schedule that allowed time "for only a cursory review."

Clearly trying to balance the contending interests objectively, the ICC staff (1987) called the Memorandum of Understanding "an innovative and appealing alternative to the traditional regulatory treatment of the . . . plants" but also said that the Memorandum contained "a variety of serious deficiencies." The staff recommended rejection.

Recognizing the adverse implications of the staff's brief, Edison quickly backtracked again and proposed several modifications to its original Memorandum of Understanding. The company continued, however, to insist on a $660 million increase. Edison's political allies praised Edison's willingness to compromise as "a major move," but consumer groups scoffed that Edison had offered only "trinket concessions" (Davidson 1987b).

On June 12 the ICC hearing examiners recommended approval of Edison's newly modified proposal with only minor changes, calling it "an innovative, appealing and unprecedented alternative to traditional ratemaking . . . [that offers] a reasonable opportunity for some rate and regulatory stability." "This sells consumers down the river," said a spokesperson for consumer groups; "It is disastrous and should be rejected" (Camper and Eissman 1987).

The examiners' draft order, and the consumer groups' reaction to it, signaled that the Commission was likely to approve the settlement. Then came what must have been a very untimely decision from Edison's point of view. "Not getting the Supreme Court case dropped is a deal-breaker," an Edison official had said in mid-December. Now—on June 16, just one day before the ICC was to begin hearing oral arguments—the Illinois Supreme Court ruled that the Commission had to reconsider the 11 percent rate increase for Byron 1. The Court shifted to Com Ed the burden of proving that the costs of building that plant had been reasonable (rather than requiring intervenors to show that the costs were unreasonable), and it found that the Commission had

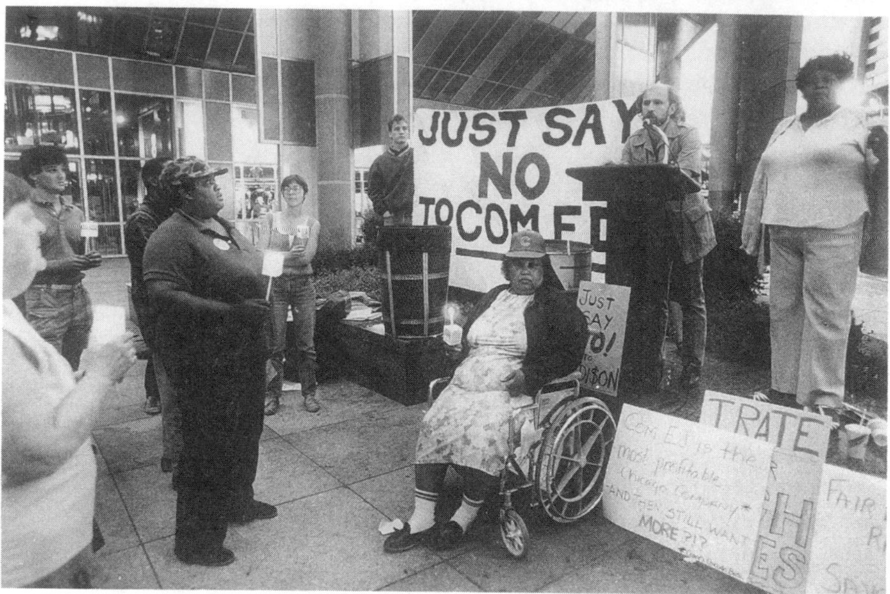

FIG. 5.4 **Opponents of Edison's Rate Increase Hold All-Night Vigil outside the State of Illinois Building in Chicago before the ICC's Hearing on June 17, 1987**

used a faulty standard when assessing the quality of the construction cost audit. This decision made the commissioners acutely aware of the need to include construction cost audits in Edison's proposed "settlement."

The Commission began hearing oral arguments on June 17. An ICC staff lawyer urged the Commission to reject the proposal unless Com Ed would include construction cost audits, but—speaking over catcalls from more than one hundred protestors who filled the Commission's hearing room—Edison's attorney replied that the utility's position was firm and that it would seek a 27 percent traditional rate increase if the Commission rejected its proposal.

Opponents of the rate increase continued to express their opposition. CUB delivered 250,000 signatures urging the ICC to reject Edison's proposal, and it asked the General Assembly to enforce the 1985 Public Utilities Act. Mayor Washington urged the ICC to require construction cost audits and restated his intent to seek alternative energy sources when the City's franchise with Edison expired in 1990.[36] A few days later the Chicago City Council voted 34–1 to oppose the "settlement."

"The company has shot itself in the foot," said Meg Bushnell during ICC discussions on July 1 (Davidson 1987c). Disturbed by the company's backpedaling over the past several weeks, she speculated that the full Commission would reject the "negotiated settlement" unless Edison made some major revisions. She and other commissioners also objected to Com Ed's unwillingness to compromise on construction cost audits, and expressed fears that the courts would overturn any decision that omitted audits. But the company remained confident: "I read between the lines very well," said one of its spokesmen, "and I can assure you that we will have a majority of votes on the commission" (Davidson 1987c).

On July 2 the ICC preliminarily voted 4–3 to reject the "settlement." Citing "fatal flaws," Bushnell said that Com Ed had never justified the need for a $660 million increase. Edison's opponents in the audience applauded each negative vote despite Bushnell's warning that the vote would not eliminate the threat of higher rates in the future. She expected Edison to file for a 27 percent increase.

A few days later, right before the Commission was to make its final decision, Edison backtracked again and offered to include construction cost audits. The ICC staff recommended approval of the revised proposal, and the Commission agreed to delay its final vote on the plan. "I believe the company is demonstrating good faith," said one commissioner who had voted against Edison a few days earlier; "My door remains open for further discussion" (Davidson 1987d). Still doubtful about whether Com Ed had justified the $660 million increase and wanting to learn whether the Commission could legally approve such an increase before construction cost audits had been completed, however, the Commission voted 5–2 to hold new oral arguments on the revised proposal. "It's a good call by the commission," said a Com Ed spokesman; "obviously we only need a change in one vote and that's what we hope the oral arguments will accomplish" (Davidson 1987e).

In oral arguments on July 14 the commissioners grilled Edison's attorneys about whether the rate increase was justified. Apparently confident of swinging one vote, Edison officials said that nothing less than $660 million would be acceptable.

Edison officials erred. On July 16 the commissioners rejected the "negotiated settlement" 4–2 with one abstention, concluding that Edison had not shown that the proposed rate increase was—as required by the Public Utilities Act—"just and reasonable." They were not persuaded by the consumer groups' charges that Edison had an unfair advantage throughout the proceedings. Finding "no indication that any improper ex parte communication took place" (ICC 1987, 491), they concluded that "while the schedule . . . has been intensive and difficult, the procedures did, in fact, afford a full disclosure of all relevant facts, and fully meets the procedural requirements of the Public Utilities Act

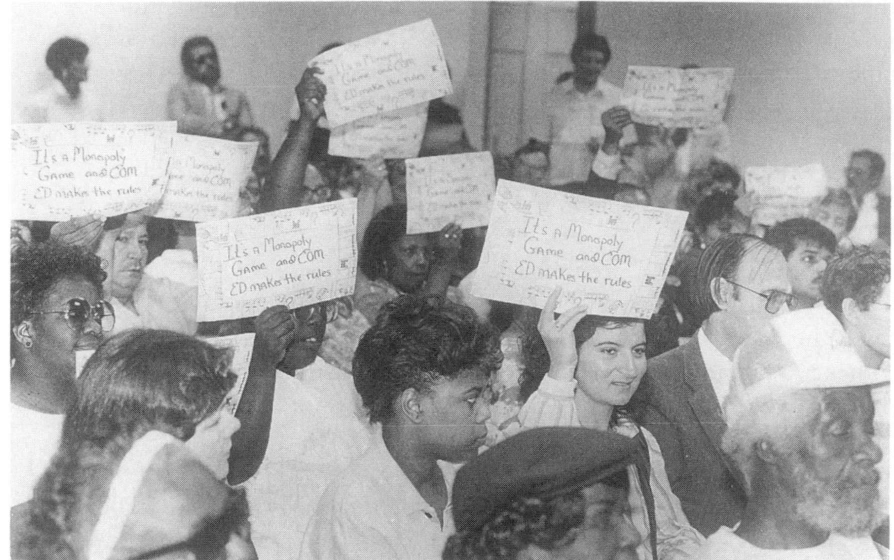

FIG. 5.5 **Consumers Protest Proposed Rate Increase on July 14, 1987**

and the constitutional mandates for due process and fair hearing" (495). Nor were the commissioners persuaded that Edison's proposal should be treated as a typical rate case. After citing relevant court cases and reviewing the evidence in the record, however, the commissioners also concluded that Edison had "not met its burden of proof as to the justness and reasonableness of the requested rate increase" (573). Though finding the proposal appealing, a majority of the Commission was obviously disturbed by Edison's "take-it-or-leave-it" approach, its insistence on an arbitrary increase of $660 million, and its numerous last-minute concessions. They wrote that

> Edison throughout these proceedings had consistently refused to support the $660 million amount in terms that would be susceptible to reasonable evaluation. . . . Thus, the Commission was basically faced with an increase determined solely by Edison on an essentially arbitrary basis and without the benefit of audits, a position which Edison refused to negotiate with any party, proponent or opponent . . . Edison's original characterization of its Proposal as a "take-it-or-leave-it" proposition shadowed the development of the entire record. (578–79)

Edison officials were stunned. "I'm simply afraid that people did not understand what we were presenting," said a company vice president;

"It was very difficult to understand precisely or even vaguely what the commissioners were thinking" (Davidson 1987f).

Unpersuasive Plots, Characters, and Point of View

I claimed at the outset of this book that good planning is a form of persuasive storytelling. In this view planners are authors of persuasive and future-oriented texts (plans, analyses, reports) that reflect awareness of differing or opposing views and that can be read and interpreted in diverse and often antagonistic ways, and that planners achieve persuasiveness by attending to the principles of good fiction writing and reader-response theory. Let us now reflect upon the extent to which the case just recounted is consistent with such a view.

If good planning is a form of persuasive *storytelling,* we must first consider how the case supports the claim that a narratival approach is superior to a nonnarratival approach. The primary alternative to a narratival approach is to think of planning as cold, hard argument; as detached and purely intellectual reasoning; as a technical process that emphasizes modeling and forecasting and sees style as purely decorative. According to this view, which Edison's managers repeatedly emphasized in public, the utility had to initiate and complete its massive nuclear expansion plan in order to meet future growth in demand, provide adequate reserve capacity, and maximize long-run economic efficiency. How rapidly would demand grow? How quickly must generating capacity additions be provided? How can the costs and benefits of those new plants best be allocated? These were all questions that could best be answered by highly skilled planners and technicians who knew how to manipulate the most sophisticated and rigorous models currently available.

Note, however, that Edison's managers had to revise their initial expansion plan time and again over a fifteen-year period. Each time they revised their plan, they were forced to take new circumstances into account. As demand growth slackened, as nuclear plants became more expensive, as consumers began to object to being required to pay for (in the consumers' view) unneeded plants, Edison's managers were compelled to speak narratively. Why did the Edison plants cost so much more than originally projected? Why were its last three units still needed in spite of dramatic reductions in demand growth? Why did the utility's December 1986 agreement and subsequent "negotiated settlement" offer the "best deal" for Illinois consumers? In each case the utility's managers had to answer by weaving a tale that emplotted the flow of action over the previous fifteen years, that identified (and frequently vilified) key characters in that flow, that showed how conflict had reached a crisis point and could be resolved in a satisfactory manner, that adopted a magisterial point of view, and that relied heavily on a scientific tone of voice to convey the truth of the company's view.

The case study shows, therefore, that Commonwealth Edison's planners and managers were storytellers who authored texts (most notably, their nuclear expansion plan, "best deal," and "negotiated settlement"). The company's plans, models, and forecasts became intelligible, meaningful, and useful only in the context of the company's continually evolving narrative as a whole. Why, for example, did the company insist that the ICC act on its negotiated settlement so quickly? In large part, because the company feared that the Illinois Supreme Court would render a decision on the Byron 1 rate case that would be adverse to the company's interest and that such a decision would drastically limit its ability to recoup its immense investment in the last three nuclear units. Timing was everything, and timing was meaningful only in narratival context.

So planning is a form of storytelling, even in an industry as technical and scientific as electric power. But it is also a form of *persuasive* storytelling. I now want to suggest that Edison's story failed to persuade consumer groups and, at a crucial moment, a majority of the ICC and that it failed primarily because its authors did not attend to the principles of good fiction writing and reader-response theory.[37] Most importantly, the story embedded in those texts had crucial weaknesses in plot, point of view, and character development.

Edison's modernist story essentially presents the company's planners and managers as heroes confronting a flawed or unheroic world. According to this plot line, Edison's planners were experts who had unique access to all the relevant information. They had aimed high and written a big plan that (1) served the interests of both consumers and investors, (2) relied on the least costly form of power generation (nuclear power), and (3) was implemented with tenacity in spite of unexpected obstacles and challenges. Despite Edison's best efforts, though, ignorant fools got in the way. Fools had pressured politicians and regulators into adopting unnecessarily costly nuclear safety standards, into delaying construction of nuclear plants, into driving the costs of those plants far above where they should be, into adopting legislation that threatened not to compensate the company for prudently incurred construction expenses, and—ah, cruelest of ironies—into not accepting a "deal" that was obviously in the fools' own best interests. Thus, Edison's heroic quest turned tragic. "Once the plan [the 'best deal'] is understood, it will be evident that it's the best deal for consumers," Edison's officials thought at first. But then, after it was all over, they realized that "people did not understand what we were presenting." Through no fault of their own, their expansion plan was proving to be another "great planning disaster" (Hall 1980).

Appealing though it may have been to Edison's officials, this plot did not persuade the consumer groups or a majority of the ICC. Most importantly, those readers saw that the plot omitted the inherent unpredictability of the future and the ability of consumer groups and the ICC to influence the price of and demand for electric power. Edison's story presumed that its planners could accurately forecast the price of and demand for electric power ten to fifteen years into the future. Furthermore, that ability to forecast future price and demand was predicated on the planners' ability to model the relationships between critical variables and to forecast changes in those variables and their interrelationships in a mathematically precise way. But consumer groups saw that Edison's forecasts did not include the actions of consumer groups, the ICC, or any other political body as a critical variable. Nor did those forecasts include other crucial unpredictabilities (such as the 1973 oil embargo and the 1979 accident at Three Mile Island). In place of expert forecasting, consumer groups and the ICC saw *ironic forecasting:* Edison's claim to have negotiated the "best deal" for consumers rested on forecasting skills that had proven severely deficient in the past.

Not persuasive in plot, Edison's story also appeared unpersuasive in point of view. The company's story presumed to be told from the central plateau of a disinterested, objective observer. Edison's planners and managers did not, however, view the electric power supply needs of northern Illinois from such a point. They were Edison employees, committed to advancing the interests of the utility and its shareholders (which they equated with the public interest). Rather than being decontextualized applications of pure logic and rigorous methodology, their models and forecasts necessarily acted as tropes in the story that Edison's management told to elected officials, consumer groups, the ICC, and the general public.

Seemingly persuasive in the late 1960s, Edison's claim to write from a disinterested point of view gradually lost credibility among many readers throughout the 1970s and 1980s. It did so, I suspect, primarily because Edison's audience sensed a growing inconsistency between the company's story (which emphasized Edison's expertise and concern for the public interest) and its actions (which revealed an inability to forecast demand and construction costs accurately; no sense of failure—let alone apology—for the wayward forecasts; and a desire to shift the costs of Edison's forecasting errors to its customers). Reeling during the 1970s and 1980s, Edison's story seemed radically inconsistent with the stance the company's managers adopted during late 1986 and early 1987: (1) their willingness to cut consumer groups out of negotiations over the "deal," (2) their dubious skill at negotiating,[38] (3) their unwillingness to provide regulators and consumer groups adequate time to analyze and discuss (hence understand) the negotiated settlement, and (4) their appearance of trying to gain an unfair advantage during the evidentiary hearings. If Edison's planners and managers were neutral, disinterested experts who had the public interest at heart, it was hard to see in their actions.

Perhaps, one might argue, it weakens my case to suggest that Edison's planners and managers really believed they were writing from such a disinterested point of view. What is really important, one might say, is that Edison clearly recognized the political nature of plan implementation and that its managers adopted the disinterested point of view as a rhetorical ploy; to complete its construction program without threatening shareholder and managerial interests, the company took a series of political steps necessary to persuade the ICC to approve the "best deal." According to this interpretation, Edison knew what was best for northern Illinois, but it had to deceive, manipulate, and trick unenlightened others into making the right decisions. They had to plan "scientifically" but act "politically."

This interpretation feels closer to the mark, and certainly reflects the position that many contemporary planners find themselves in. Before accepting this interpretation, though, one must ask why Edison—despite the vast amount of resources available to it—still failed to persuade the commissioners. I want to suggest that Edison failed to persuade the ICC commissioners precisely because Edison's planners and managers did not read and critique the consumer groups' story carefully and then revise the company's initial story in light of that critique. Stated differently, they failed to persuade a majority of the ICC to approve the "best deal" because critical readers did not find Edison's plot and characterizations to be believable.

That Edison continued to believe that its story deserved privileged status (and that it failed to read and critique the consumers' story carefully) can be seen in its characterization of consumers and consumer advocates. According to Edison's nuclear power expansion plan and associated forecasts and models, consumers were simply economic agents (rational, self-interested, utility-maximizing individuals) whose behavior could be monitored and controlled from an objective, scientific, decontextualized point of view that was "above" the political fray:

"Once the plan is fully understood," an Edison official said, "*it will be evident* that it's the best deal for consumers" (emphasis added). But the context of decision making compelled Edison officials to write and use those methodologies in a consciously persuasive way (as tropes), and the tragedy they recounted (and their actions with regard to the "best deal" and the negotiated settlement) treated those consumers as meddling fools. "Obviously we only need a change in one vote," they said, "and that's what we hope the oral arguments will accomplish."

According to Edison's characterization, therefore, consumers were simply "rational, self-interested individuals" and consumer advocates were nothing but "meddling fools." Such a characterization was radically at odds with the consumer groups' self-conception. When consumer advocates read Edison's texts they did not see the company's planners engaged in an objective search for the "one best way" of acting in the "public interest." They saw a profit-motivated nuclear construction program that asserted itself with ever-growing insistency and that sought to violate the 1985 Public Utilities Act no matter how bad its consequences might be for consumers. They did not see themselves simply as rational, self-interested individuals. Rather, they thought of themselves as a group of dedicated people (both expert and lay) who were engaged in a romantic quest to overcome evil. Edison's "negotiated settlement" was, in their view, "a mine field for consumers . . . smoke and mirrors . . . a clever scheme . . . [that] sells consumers down the river." No matter how closely they read, they could not find themselves or their stories fairly characterized in Edison's texts. And, by presenting radically impoverished characterizations of those groups and their arguments, Edison lost the opportunity to reconstruct a more persuasive story.

Edison's planners and managers failed, therefore, to persuade a majority of the ICC because they did not persuasively respond to the antagonistic reading embedded in the consumer groups' story. Consumer groups, on the other hand, succeeded in persuading the majority. Why? They succeeded in large part because their story emplotted the flow of action clearly and consistently: consumer groups were engaged in a romantic quest to defeat a grossly mismanaged firm that was trying to bloat its profits dramatically by manipulating elected officials, obtaining unfair advantages in the ICC's hearings, presenting the commissioners with a "take-it-or-leave-it" proposition, and reorganizing its way around the Public Utilities Act of 1985. Furthermore, the consumer groups told their story from a consistent and believable point of view: they claimed simply to represent the interests of consumers. In their view, it was in the interest of consumers for Edison to demonstrate—through construction audits as required by law—that the costs of building its last three units were "just and reasonable." In this context, the Supreme Court's decision on the Byron 1 case acted as an extremely powerful trope for the consumer groups' story. That decision made it quite clear to the commissioners that their own decision had better conform to the Public Utilities Act or the Court would overturn it.

Perhaps the most notable weakness in the consumers' story concerns the tropal nature of demand forecasting. Consumer groups vigorously critiqued Edison's forecasts, pointing out the irony of relying on techniques that had proved so far off the mark in the past. But, if Edison's forecasts were so erroneous, why should the Commission be persuaded that the consumer groups' forecasts would be any more accurate? Here I can only point out that forecasts and other techniques act as tropes in planning arguments and that planners need to begin discussing ways to evaluate them as persuasive figures of argument rather than as decontextualized applications of cold, hard logic.

I do not want to leave my case study at that point, however. It is important to see the constraints within which future-oriented storytelling and critical reading take place. The meaning of Edison's plans, forecasts, and deals did not lie simply in Edison's intentions (whatever they might have been). Nor did that meaning reside simply in Edison's texts themselves. The meaning of those plans, forecasts, and deals also lay in what the consumer groups read into those texts. Does this mean, then, that a plan, forecast, survey—even a "negotiated settlement"—can mean whatever readers want it to mean, that planning is an entirely relativistic enterprise? I think not. Edison, its governmental allies, the consumer groups, and the ICC can all be seen as part of an "interpretive community" (Fish 1979) that shaped how they wrote and how they read. Engaged in the "normal discourse" (Rorty 1979) of that community, Edison and the others argued in a context of prior experiences with one another, shared understandings, and a common language. As shown in chapters 3 and 4, they knew the prior ICC and court decisions that bounded the commissioners' discretion, they agreed that Edison's nuclear plan had resulted in generating supply that (through the late eighties at least) substantially exceeded demand, and they talked in terms of rate base, rate of return, present value revenue requirements, petitioners and intervenors, briefs and rebuttals, and excess capacity. Though opponents, they all tried to persuade the ICC commissioners in terms of the "agreed-upon conventions" of public utility planning and regulation. Planning is persuasive storytelling about the future, but the quality of that storytelling is continually measured in terms of the standards applicable to the localized community within which that planning takes place. What happens when planning occurs in a context of an "abnormal discourse" between interpretive communities—when the community and its institutions are at stake—is a topic that I will return to with vigor in chapters 7 through 9.

I have claimed that planners are authors of persuasive and future-oriented texts (plans, analyses, reports) that reflect awareness of differing or opposing views and that can be read (constructed and interpreted) in diverse and often conflicting ways. Edison's planners and managers authored such texts (their nuclear expansion plan and subsequent "best deal" for consumers), but their texts failed to recognize that they could be read in antagonistic ways. I have also claimed that planners achieve persuasiveness by following key principles of good fiction writing and reader-response theory. Edison's planners and managers failed to persuade consumer groups and the ICC because the company's story was weak in plot development, written from an unbelievable point of view, and characterized consumer groups in a way those groups could not identify with. If this interpretation strikes the mark, then planners need to pay far more attention to how they write and read their texts and to the contexts within which that writing and reading takes place. Reader response makes their texts "political" despite any modernist pretensions to the contrary. Planners also need to learn how their technical skills (forecasting, surveying, modeling) act as persuasive imagery within their texts and to learn why those tropes help to persuade some audiences but not others. Last, planners and other professional experts need—as I hope this chapter shows—to discover the pleasure and pain of writing and interpreting richly complex and enlightening cases.

Interlude

The Illinois Supreme Court's Chambers in Springfield

I walk down the street from the ICC's office to the State Supreme Court building. It is a stately three-story structure, built in the neoclassical style popular in the later 1800s. I climb the stone stairs, enter the building, and turn left toward the the chief clerk's offices. I am struck by the contrast with the ICC's offices. The clerk's offices are—unlike those at the ICC—clean, uncluttered, and rather handsome. The staff likewise strike me as being younger and more becoming. It does not feel like a place to end up, but like a place close to important action. I go upstairs to the court's chambers. The courtroom is large, stately, and impressive—really quite different from the ICC's main hearing room. The seven justices, all older white men, sit behind a large, curvilinear, and elevated wooden desk. A podium stands in front of the desk, facing it, enabling the counsel to speak at eye level to the justices. Seventy or so darkly wooded and plushly black-leathered chairs fill the room. The room itself is ornate, with four dark red curtains, a red carpet, pick lamps on the wooden walls, gold-plated crowns on pillars, romanesque murals high on the walls, and a lavish mural on the ceiling. Each mural expresses some aspect of the law. The caption of one reads "Law destroys violence [and] anarchy"; of another "Law exposes fraud [and] discord"; and of a third "Law protects industry [and] peace." At the rear of the room, above the door, are the words "Audi Alteram Partem." Hear the other side.

We're glad to have two years of regulatory battles behind us.

— A COM ED OFFICIAL

[The decison] reflects an illegal settlement between the Commission and Edison.

— ILLINOIS SUPREME COURT

Edison Completes Its Nuclear Power Expansion Plan, But Who Will Pay for the Last of It?

Edison applied for ICC approval of a "negotiated agreement" and Memorandum of Understanding. It would bind the company, Governor Thompson, Attorney General Hartigan, Cook County State's Attorney Richard Daley, and a few other minor organizations. The agreement was, according to its proponents, a response to the Commission's suggestion to explore alternatives to traditional ratemaking. It would (1) transfer Edison's last three nuclear units to a wholly owned subsidiary of Edison, (2) entitle Edison to purchase all power generated by the subsidiary for the first five years of operation, (3) give the company a $660 million rate increase (coupled with a five-year moratorium on future rate increases), and (4) resolve other related cases pending before the Commission (ICC 1987). Consumer groups, the City of Chicago, and others strongly opposed the negotiated agreement, however, and the Commission ultimately rejected it by a 4–3 vote in early July 1987. In deciding not to approve the proposal, the Commission noted that

> the Commission was basically faced with an increase determined solely by Edison on an essentially arbitrary basis and without the benefit of audits, a position which Edison refused to negotiate with any party, proponent or opponent. . . . [But] the Commission again wishes to emphasize strongly that it encourages negotiation and innovative approaches to resolving the issues which Edison's Proposal sought to address. By this Order the Commission has not rejected the basic concepts contained in the Proposal. . . . However, any such proposal must satisfy statutory requirements, provide the Commission with the ability to determine the justness and reasonableness of the

resultant rates, and be supported by an evidentiary record which permits the Commission to conclude that the proposal is in the public interest. (ICC 1987, 578–79)

The Supreme Court's ruling concerning the Byron 1 rate increase deeply influenced the commissioners and greatly increased the persuasiveness of the consumer groups' arguments. The key issue in the case was whether the costs Edison incurred in building that unit had been "reasonable." Noting that the 1985 amendments to the Public Utility Act gave Edison the burden of proving that its costs were reasonable, the Court concluded that the Commission had improperly placed the burden of proof with the intervenors to show that Edison's costs were *un*reasonable. Noting also that the 1985 law required the audit to be conducted in accordance with "generally accepted auditing standards," the Court concluded that the Commission's reduction of that standard to "professional competence" was insufficient. These two deficiences persuaded the Court to instruct the Commission to reconsider its decision on Byron 1. But the Court also allowed Edison to continue charging the rates approved by the Commission in 1985 until such time as the Commission altered its decision and required Edison to repay consumers for any excessive rates (Illinois Supreme Court 1987).

Edison's effort to have consumers pay for its last three nuclear units had, for the most part, failed. At least in this instance. But someone would have to pay for them. Construction of Byron 2 and Braidwood 1 had been completed. Byron 2 was being placed in service in April 1987, and Braidwood 1 would soon follow. Braidwood 2 was roughly 85 percent complete. So, stung by the Commission's rejection but wanting to recoup the cost of its investment, Com Ed applied for a traditional 26.9 percent, or $1.41 billion, rate increase on August 21, 1987. It also asked the ICC to reconsider its rejection of the prior proposal. ICC Chairwoman Bushnell worried out loud that the simultaneous requests were an effort to pressure the ICC into approving at least one, but Com Ed Chairman O'Connor replied that Edison had asked for a rehearing in order to focus on core issues that got lost in the discussion of legalities. Governor Thompson, State's Attorney Daley, and consumer groups attacked the request for a 27 percent increase in rates, with Thompson condemning it as "absolutely unwarranted." "We will fight to the death against a 27 percent rate increase," he said. On the other hand, Commissioner Manshio, Com Ed's strongest supporter on the ICC, found the application to be "staggering but a realistic reflection of the whole situation" (Davidson 1987g).

"The whole situation": though Commissioner Manshio could not know it at the time, the whole situation would soon become far more complex and confusing. The Commissioners, Edison's management, consumer groups, and others were becoming lost in a maze of ever-thickening complexity. Within the next two and a half years, the Commission approved a "settlement agreement" pertaining to the 27 percent request, only to have that decision overturned by the Illinois Supreme Court in late 1989. While attempting to deal with that request for a rate increase, the Commission also had to respond to the Supreme Court's June 1987 decision about Byron 1. Eventually the ICC ordered the company to refund $190 million to its customers and to reduce its rates by $43 million. Edison's rates bobbed like a cork in the ocean, surging upwards with ICC decisions then back down following court appeals and remands. And while Edison's rates bobbed, the City of Chicago continued to explore alternatives to remaining on the company's system. By the end of 1989, the City had decided to terminate its franchise agreement with the company. The modernist story had no room whatsoever for such developments; other authors had taken over and begun writing a far more complicated tale.

The Commission Staff Proposes a "Settlement Agreement"

To find a path through the maze of ever-thickening complexity, it would help to focus first on the company's renewed effort to obtain a rate increase for its last three plants.[39]

In early September, the ICC rejected (4–3) Edison's request to reconsider the July decision, and its hearing examiner began dealing with the request for a 27 percent increase. The case was complicated by the fact that audits of the last three units had not yet been completed, so the hearing examiners held evidentiary hearings on the non-audit portions of the rate increase from January through mid-June 1988.

While the hearing examiners were collecting testimony and evidence, Edison and the intervenors were involved in renewed efforts to negotiate a new, nontraditional settlement acceptable to all, or at least most, of the major participants. Shortly after the company had submitted its August request for a rate increase, Chairwoman Bushnell had urged Com Ed to continue negotiations with consumer groups in an effort to reach agreement with at least some of them. Though I am not privy to any detailed records of the negotiations that followed, my sense is that they formed two stages: the first in late 1987 and early 1988, the second in mid-1988.

The major elected officials involved in the previously unsuccessful effort to negotiate a resolution to the case had good reason to respond positively to Bushnells' advice. The Commission's July decision had embarrassed Thompson, Hartigan, and Daley, and successful negotiations might remove the stain. In late January 1988, Governor Thompson tried to revive the original "best deal." Recalling the previous embarrassment, though, he also suggested that any new deal should "leave more money in the consumer's pocket." Com Ed management responded with some interest, but BPI and CUB denounced the suggestion. "The timing is extremely odd," said Susan Stewart, and Douglas Cassel raised the possibility that "it could be an effort to sidetrack the discussions that are currently underway" (Dold 1988a).

Private negotiations between Edison and intervenors continued through early February 1988. One meeting happened just a few days before the ICC staff and others were expected to call for significant reductions in the utility's rates. (For example, the Illinois Industrial Electric Consumers Association [IIECA], which represents about twenty corporations responsible for the purchase of about one-fifth of Com Ed's total industrial sales, was going to call for a reduction of about $300 million.) The meeting was supposed to be secret, but reporters learned about it. The gathering included representatives from Com Ed, CUB, BPI, IIECA, the City of Chicago, Cook County, the Office of the Illinois Public Counsel, the Illinois Small Business Utility Advocate, and the offices of Thompson, Hartigan, and Daley: twenty-six people altogether. Pleased by the magnitude of rate reductions being recommended by the industrial consumers, and feeling that the IIECA's proposal weakened Edison's hand, CUB and BPI welcomed the talks. An Edison spokesman said simply that "everything is negotiable" and that the company "would potentially be willing to settle for less than $660 million as part of a complete package" (Davidson 1988a).

A few days later, the industrial group argued in the ICC's evidentiary hearings that shareholders should not be allowed to earn a return on investment in the three new plants until they were used and useful. Prepared by Drazen-Brubaker and Associates, a Saint Louis utility consulting firm, the industrial group's report contended that Com Ed's rates would be near the highest in the country if the 27 percent increase was granted. "Edison is in real trouble now," said Stephen Moore, Illinois public counsel. "The numbers just do not justify the rate increase they are asking for" (Davidson 1988b).

In mid-February, the *Chicago Tribune* reported that consumer groups had met with political and industrial representatives and with ICC staff to share data that suggested the utility might be in line for substantial rate cuts rather than an increase. They had met on January 20. Com Ed, which had been excluded from the meeting, expressed interest in continuing the negotiations. BPI signaled that it might be willing to accept a small increase. As Cassel put it, "settlement is vastly preferable to litigation, and we don't want to say to Edison 'It's zero increase or no deal.' . . . We might even be willing to consider a small increase" (Davidson 1988c). Gradually, it seemed, the contending parties were moving closer together.

In early April, the ICC staff recommended that Edison's rates be reduced by $343 million for about six months and then be increased by $235 million. Further increases would be phased in over the next few years (Com Ed 1989). Perhaps a settlement was near. Time was running out on the Commission, though, and its hearing examiners had not yet had the opportunity to hold hearings on the audits of Byron 2 and Braidwood 1.

In mid-April, the *Tribune* reported that the audit of Braidwood 1 was expected to show wastes of between $100 and $600 million, depending on the method used. Com Ed representatives quickly visited the *Tribune* and other papers to defuse the adverse effects that they expected this announcement to produce. A spokesman for the company said it was concerned that "the auditing hierarchy has recently changed the approach to arrive at a more punitive number . . . It's inexplicable" (Davidson 1988d). Consumer groups publicly wondered how Edison had gained access to the audit before it was released.

The audits of Braidwood 1 and Byron 2 were formally released on April 14 and April 18 respectively. Prepared by O'Brien, Kreitzberg and Arthur Young, respectively, these audits could not have been good news for the company. Or for its shareholders. The Braidwood 1 audit concluded that between $264.4 million and $278.3 million (8.1 to 8.5 percent of the unit's total cost) should be excluded from Edison's rate base. According to the company's 1988 annual report to shareholders, the auditors claimed that

> the unreasonable costs were primarily the result of an aggregate twenty-nine month delay in the completion of the generating unit, which delay they attributed to (i) the Company's financing practices and overdependence on uninterrupted financial support through timely rate relief, both of which they asserted led to a cash shortage in late 1979 that resulted in the shutdown of construction from late 1979 until April 1980 and slowed construction from April 1980 to late 1981, and (ii) a failure to develop adequate construction control procedures to assure adequate quality and demonstration of that quality which they asserted resulted in increased reinspection, reassessment and rework programs that resulted in delays in construction and delays to the issuance of an operating license by the NRC for the unit. (Com Ed 1988, 31–32)

Braidwood 1 would have cost $2,396 million instead of $3,268 million had it not been for these "unreasonable" costs.

The company's commentary for the Byron 2 unit was virtually identical, with only the data differing. Thus Arthur Young had concluded that between $119 and $181 million (6.3 to 9.5 percent of the approximate total cost of $1,899 million was unreasonable and should not be included in the company's rate base. These costs were principally due to an aggregate 23.6 month delay. If not for those delays and costs, the cost of Byron 2 would have been approximately $1,520 million instead of $1,899 (a $379 million difference).

Edison characterized O'Brien, Kreitzberg as being particularly tough on utilities, and it asked the ICC to investigate the possibility of improper communications between ICC staff members and consumer groups.[40] Then it formally disputed the recommendations of the auditors in both instances.

Not many days later, the company received some more bad news. On April 28, the re-audit of Byron 1 was publicly released. This audit, again conducted by Arthur Young, concluded that between $134 and $169 million of the $2,537 million cost of building that unit had been unreasonable, the result of an aggregate construction delay of 15.9 months. Once again, the unreasonable costs were attributed to project management, cost monitoring, and quality assurance problems. Another $232 million excess cost would not have been incurred had it not been for the construction delays (Com Ed 1988, 33).

What to do? The audits had been completed (except for Braidwood 2), but the Commission did not have enough time to conduct evidentiary hearings on them and still comply with the act's deadlines. But if they did not act now, the 27 percent rate increase that Edison had requested back in August would take effect automatically on July 17 (Com Ed 1989). The *Chicago Tribune* suggested one possible solution. In an editorial on May 1, it argued that "open warfare" between Com Ed and the ICC needed to be "settled quickly before the region's economy is severely damaged," and that what the people of the region need is "a persuasive leader or group not caught up in the front-line combat who can put a reasonable new proposal on the table. . . . ICC's staff of experts and technicians," the *Tribune* said, " . . . must shun politics and propose a sensible compromise" (*Chicago Tribune* 1988b).

Just a few days later, the ICC staff asked the Commission to defer further consideration of the proposed rate increase (for which a proposed order was expected imminently) and to authorize the staff's Executive Director to negotiate a settlement between Edison and the

Commission (Stone 1988). The proposed settlement would allow Edison to increase its rates by $235 million effective January 1, 1989, and by $245 million one year later, then would freeze rates through December 31, 1993. The Commission was attracted to the proposal, but—aware that the Public Utilities Act gave them only a few more weeks to decide on Edison's request for a rate increase—indicated that Edison would have to withdraw and refile its proposed rate increase if it wanted to negotiate with the Commission as the staff suggested. A company spokesman said that Edison was "gravely disappointed in both the amount and timing" of the increase. Douglas Cassel of BPI and Susan Stewart of CUB did not like the staff's proposal either: in Cassel's view, the Commission had proposed "to give away the store"; Stewart called the proposal "outrageous" (Unger 1988).

A few days later, the company withdrew and refiled its request for a 27 percent increase, and it agreed to discuss the staff's proposed settlement. Consumer groups and the Illinois public counsel denounced it as a "sham" and a "sweetheart deal," wondering how the staff could justify the 8.2 percent figure on the record. The staff then held private conferences with all interested parties over the next month and a half. In the end, however, it concluded that not all parties would agree to the proposed settlement (Com Ed 1989).

By early August, the ICC staff had revised its 8.2 percent proposal and had submitted a formal "Offer of Settlement." The revisions were modest and left consumer groups far from satisfied. One modification concerned fuel savings: if Com Ed ran its nuclear units so efficiently that it saved more than $306 million in fuel (by not using fossil-fuel plants), then it would split the savings evenly with customers. The earlier proposal had all the savings go to the customers. Similarly, the new proposal said that Com Ed could net fifty cents for every dollar it earned by selling electricity to other utilities beyond the first $21.2 million. Earlier Com Ed could net only

Recall the *regulatory bargain* and the notion that the ICC was to *balance the interests* of shareholders and consumers. Suddenly the Commission's staff begins acting as a neutral arbiter in negotiations between Edison and the Commission. In what sense is this consistent with the modernist idea that the Commission is supposed to balance interests? Confusion about *balancing interests* deepens. In an editorial of June 15, the *Chicago Tribune* praised the ICC staff for its initiative. The editorialist wrote that the situation was analogous to a boxing match with Com Ed in one corner, the ICC in another, angry customers seated at ringside, and the staff ending the fight. This is a very odd analogy. Given the legal task of balancing interests, shouldn't Com Ed and its customers be in opposite corners, with the Commission acting as referee?

twenty cents. Consumer groups were upset. "What they've got here is a unilateral proposal," said Stewart. "It's not a settlement, yet all of a sudden we're supposed to put our two cents in. This really undermines public confidence in how utility rates are set here. . . . We're going to court" (Biddle 1988). Edison's spokesman, on the other hand, indicated that the company was satisfied with the process, but thought that the price was too low. A few days later, on August 5, 1988, Edison completed the construction of the Braidwood 2 unit and placed it into service. Only eight years behind schedule.

From mid-September through mid-October, the hearing examiners held evidentiary hearings on the Offer of Settlement and the Byron 2 and Braidwood 1 audits. In mid-September, CUB argued that the ICC should reduce Com Ed's rates by at least $413 million rather than increase them by $480 million. In mid-November, ICC hearing examiners urged the Commission to authorize a $235 million increase effective January 1, 1989. This increase accounted for completion of Byron 2 and Braidwood 1 and the subsequent audits of the reasonableness of their costs, but the examiners felt that they could not include the effects of Braidwood 2 until the unit's audit had been completed.

In early December 1988, Com Ed proposed a two-step, $795 million rate increase during oral arguments before the ICC: $235 million followed by $560 million a year later. This was equivalent to its 1987 request for $660 million. CUB, BPI, the City, the Cook County state's attorney, and others argued that Com Ed had not proven the need for the three plants. Any rate increases, said Cassel, would be like "laying before Edison a whole Christmas tree surrounded by presents" (Karwath 1988a).

The ICC hearing examiners recommended $235 million followed by $245 million, accompanied by a freeze through 1993. The second figure was contingent on the results of the Braidwood 2 audit.

On December 21, 1988, the ICC stated its intent to approve the staff's proposal, with modifications, as just and reasonable if Edison agreed to the conditions. It tentatively authorized Edison to increase its rates by $235 million effective January 1, 1989, and by $245 million one year later (depending on the results of the Braidwood 2 audit and other matters). The decision imposed a five-year moratorium on requests for a general rate increase (subject to certain contingencies), and authorized customer refunds for any of the years from 1989 through 1993 (under certain

Who will pay the costs of decommissioning Edison's nuclear plants when their "useful lives" (twenty-five to thirty-five years) are over? In 1988, Edison reported that it had (in compliance with an NRC order) created and would fund "external trust

circumstances). It also guaranteed customers $306 million fuel cost savings (while retaining the equal split of cost savings above that amount), terminated the case concerning 1986 federal income tax changes, and ensured completion of the Byron 1 case. Com Ed could opt out of the plan in 1990 if it felt the second increase was not high enough, but the ICC could then begin hearings on whether to refund any of the first step. Consumer groups opposed it. Protesters carried signs saying "Another ICC Christmas. Take care of the greedy. Take from the needy." Commissioner Susan Stone, arguably the most pro-consumer member of the ICC, said that she "drew up a balance sheet and . . . that the benefits outweigh the negatives in the final outcome" (Karwath 1988b).

funds" to cover the costs of decommissioning. It estimated that decommissioning costs were approximately equivalent to the difference between the 3.50 percent annual depreciation rate for the years 1989 through 1993 and the 4.05 percent annual rate for 1988, or $2.2 billion in current-year dollars for all of the Company's nuclear units.[41]

On December 23, Com Ed's Board approved the rate plan. "We're glad to have two years of regulatory battles behind us," said a company spokesman (Karwath 1988c). Consumer groups said they would appeal to the Appellate Court on the grounds that CUB, BPI, and other intervenors had been compelled to participate in "precipitously scheduled 'settlement' hearings which: (1) were subject to no written or stated

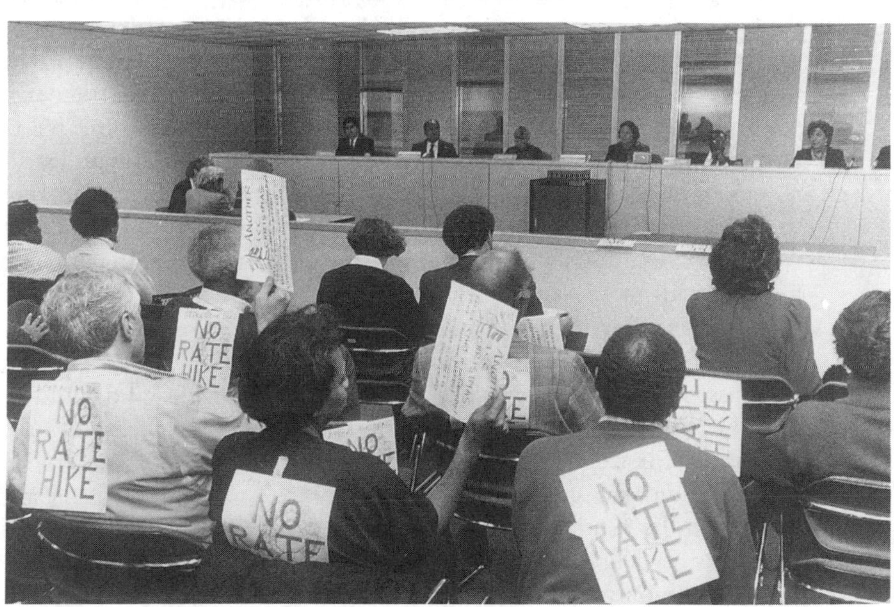

FIG. 6.1 **Consumers Protest Edison's Proposed Rate Increase on December 21, 1988**

rules; (2) were subject to new burdens of proof articulated by the Examiners for the first time in these hearings; (3) proceeded on an accelerated schedule which did not allow for adequate preparation, development of necessary evidence, or complete presentation or examination of the 66 expert witnesses; and finally, (4) had a preordained result based on the proposed deal rather than on the merits of the evidence presented, evaluated under the statutory standards mandated by the Act" (BPI/CUB 1989b, 87). Four days later, the City of Chicago and the Office of Public Counsel sued the ICC in the Cook County Circuit Court to prevent the Commission from voting on the plan, claiming that the rate increase was a "bargain," not a finding based on need and costs, and hence that it violated the Public Utilities Act. The Cook County State's Attorney's office asked the ICC to reconsider the plan. On December 28, the Circuit Court temporarily restrained the ICC from voting on the rate increase. The judge expressed doubt that Com Ed would be able to refund all the money collected through higher rates, plus interest, if the rate increase was later declared illegal. The ICC and Com Ed asked the Appellate Court to stay that judge's ruling. It did so.

On December 30, the ICC unanimously approved the rate increase (which came to be known as the Sixth Interim Order), though Commissioners Stone and Foran expressed several reservations (ICC 1988). In doing so, the Commission noted that some intervenors claimed that the Offer of Settlement was not really a settlement: the settlement process had failed, the parties had not agreed, the staff had filed a unilateral offer of settlement, and the standards of review for settlements were not applicable to the staff's offer. In response, the Commission stressed that the staff's Offer of Settlement "is not a settlement agreement and should not be judged by the standards for such an agreement. . . . [Rather] Staff presents its Offer of Settlement as a just and reasonable resolution of the issues in the subject proceedings on the merit, based on substantial evidence in the record as a whole. Staff proposes that its Offer of Settlement be reviewed pursuant to traditional just and reasonable standards" (ICC 1988, 6).

In assessing what construction costs were "reasonable" and could be placed in Edison's rate base, the Commission explicitly addressed three key issues. First, it had to decide whether the audits complied with applicable auditing standards. Here the Commission noted that the audits had to be conducted in accord with generally accepted auditing standards as delineated in a U.S. General Accounting Office publication commonly called the "Yellow Book." The Commission also stressed that "the nature, timing and extent of the procedures to be applied are based on professional judgment. The amounts and kinds of evidentiary matter required to support informed opinions or conclusions are matters for auditors to determine after careful study of specific circumstances. Further, the auditors' opinions or conclusions must be formed within a reasonable length of time at a reasonable cost to be economically useful" (ICC 1988, 14). Second, the Commission had to assess whether the auditors used proper methodologies for conducting their audits. And third, it had to assess the merits of a series of specific conclusions regarding the delays and costs identified by the audits. The Commission found that "no affirmative evidence was offered to impeach the evidence that . . . it was prudent and reasonable for Edison to conclude that Byron Unit 2 and Braidwood Unit 1 would be used and useful in providing service to its customers" (24). The Commission also accepted testimony from well-known experts (including two Nobel Prize economists—George Stigler and Kenneth J. Arrow—who testified on Edison's behalf) about the meaning and implications of the "used and useful" standard. It concluded that Byron 2 and Braidwood 1 were used and useful within the meaning of the Public Utilities Act.

Commissioner Stone did not entirely agree with the decision. In a separate statement, she expressed considerable frustration with the final order's vacillation between treating the rate increase as a traditional case and treating it as a settlement. She also disagreed with the processes which led to the order, particularly the role of the staff during the negotiations:

> This Commission has encouraged negotiation among parties and has approved a number of successfully stipulated settlements in the recent past. . . . In those cases, however, others took the initiative, advanced and argued the proposals, while Staff's role was to impartially examine the facts, analyze the consequences, suggest changes and recommend action by the Commission.
>
> By contrast, in this case, Staff acted as prime advocate. Having authored the proposal, albeit with comment from other parties, Staff quite understandably appeared wedded to its major components. . . . This resulted in reciprocal rancor, suspicion and ill will. Many parties, including BPI, CUB, Chicago, OPC, LIRC and others, who felt frozen out of making meaningful modification, then withdrew from negotiations.
>
> The rate base treatment of Edison's three nuclear units presents far more than a normally contentious case. Thus, regardless of the merits of Staff's carefully crafted package or of the assumptions on which its components rested, the credibility of the negotiation process, as implemented, was at serious risk. The process was ill advised and failed to secure the desired critical

> mass of consensus. Thus, without the sought-after broad support by the August 1 deadline, it became necessary to shift to considering the components of Staff's Offer as an integrated package within a regularly litigated rate case context. (Stone 1988, 3–4)

On January 3, 1989, a divided Illinois Appellate Court refused to block the rate increase temporarily. As promised, the several intervenors (including BPI, CUB, the City of Chicago, the Cook County State's Attorney, the Office of Public Counsel, and others) who did not agree to the Commission's Sixth Order quickly appealed the order directly to the State Supreme Court. Later in the month, the Supreme Court agreed to hear the case.

On February 28, 1989, the ICC released the audit of Braidwood 2. Once again prepared by O'Brien, Kreitzberg and Associates, the audit concluded that the decision to complete the unit had been reasonable but that construction of the unit had been unreasonably delayed by 26.6 months. Consequently, between $161 and $195 million of the unit's $1.86 billion construction cost had been unreasonable. Another $260 million in costs would not have been incurred, had it not been for the unreasonable delay (cited in Com Ed 1989, 40). The Commission would have to decide what portion of Braidwood 2's cost to include in Edison's rate base and what portion to exclude on the grounds that it had not been prudently incurred, had not been reasonable, or was not used and useful.

Deciding How to Respond to the Illinois Supreme Court's 1987 Ruling on Byron 1

By the middle of 1989, it appeared as though the Commission had finally decided who should pay for Edison's last three nuclear units. With that issue out of the way, it could now reconsider what to do about Byron 1. Recall that the Supreme Court had ruled in mid-1987 that the ICC improperly decided what portion of that unit's costs had been reasonable. The Court sent the case back to the ICC with instructions that the Commission reconsider the reasonableness of those costs and order refunds to consumers for any excess rates.

To help it respond to the Court's order, the Commission had ordered a second audit of the unit. Conducted by Arthur Young, this second audit was publicly released in April 1988. It concluded that between $134 and $169 million (or 5.4 to 6.8 percent) of the unit's $2,537 million total cost could be deemed unreasonable. An aggregate construction delay of 15.9 months caused Byron 1's costs to be $409 million greater than they would otherwise have been (Com Ed 1989). While

Arthur Young was conducting its audit, Com Ed was collecting bill payments from its customers, payments that reflected the Byron 1 rates originally approved back in October 1985.

Distracted by Byron 2 and the two Braidwood units, the Commission was unable to focus on the Byron 1 refund until mid–1989. It was an important case; large amounts of money were involved, it would affect Com Ed's write-offs and rate base for the next thirty years, and it might establish precedent for how the Commission would handle future remand cases concerning Edison's last three units. (Howard Learner of CUB suggested that it was "probably the most significant utility rate case of the last decade" [Karwath 1989a].) It was also a difficult case to resolve; and a majority had to be forged out of an even number of commissioners because Commissioner Foran had removed himself from the case. Compounding the difficulty was that the Supreme Court had forced the Commission's hand by imposing a deadline for Commission action.

At first, Commissioners Bushnell, Stone, and Romero voted to find that $409 million of the costs had been unreasonable, whereas Commissioners Barrett, Manshio, and Kretschmer argued for as little as $134 million. Barrett appeared to be the swing vote, and it was unclear what it would take to move him one way or the other. "What it really gets to," Learner said, "is how much of the store do Stone, Bushnell, and Romero have to give away to get Barrett to go along" (Karwath 1989b). The Supreme Court extended its deadline another two weeks to give the commissioners time to construct a majority.

On August 9, the ICC tentatively decided 5–0 (with Foran removed and Kretschmer not present) that approximately $200 million of Byron 1's construction costs had been unreasonable, principally because of the construction delays, and that those costs should be excluded from Edison's rate base. The company would also have to refund approximately $190 million to its customers. An Edison spokesman said that the utility was "shocked and very disappointed" by the Commission's tentative decision and that the company believed "a disallowance of this magnitude is ex-

FIG. 6.2 **ICC Chair Bushnell**

cessive and unfair." Conversely, consumer groups thought the refund was not large enough. "The driving force behind this order now is to reach a deal and then toss the legal and factual issues to the court for review," Howard Learner said (McRoberts 1989). BPI intended to seek immediate implementation of the Commission's order but also to appeal on the grounds that the refund should have been about $400 million.

On August 23, the ICC finalized its $190 million refund order by a 4–2 vote. Predictably, Com Ed and various intervenors appealed the decision. Two months later, on October 31, Circuit Court Judge Richard Curry ruled that Com Ed should refund not $190 million but $248 million to account for interest that customers could have earned on that money had they not had to pay it to the company. The company appealed that ruling to the Appellate Court, which is where the case was at the end of 1989. In that same month, the company "recorded" some of the effects of the Byron 1 disallowances and rate refund with interest; doing so reduced Edison's net income by approximately $62 million or $0.29 per common share. It was clear that the company would have to record additional effects early in 1990 (Com Ed Annual Report 1990).

On December 22, the ICC ordered Edison to reduce its rates by an additional $43 million, effective January 1 of the following year, based on the additional disallowance of the $200 million unreasonable costs at Byron 1. The company appealed the decision and continued to collect revenues on that $200 million pending the outcome of the various appeals that were currently underway.

The State Supreme Court Overturns Another ICC Decision

On December 21, just the day before the Commission ordered the $43 million rate reduction, the Supreme Court overturned the 1988 two-step rate increase for Byron 2 and Braidwood. The main issue, in the Court's view, was whether the Commission had the legal authority to issue the 1988 order. Despite the Commission's claim that the 1988 decision was not a settlement agreement, the Court concluded that "the Commission used a settlement rationale to justify the Sixth Order. The Commission had no authority to impose a settlement not agreed to by all of the parties and the intervenors. Consequently the Commission improperly entered into a settlement with Edison. Moreover the Commission did not base its decision in the Sixth Order exclusively on the evidence presented in the record" (Illinois Supreme Court 1989, 15). The 1988 decision, the Court ruled, "reflects an illegal settlement between the Commission and Edison. The Commission went beyond the scope of its authority . . . by entering the Sixth Order without the agreement of the intervenors. In addition, the manner in which the Commission decided the Sixth Order violated the Act to the prejudice of the intervenors. . . . We also hold that the Sixth Order was not 'supported by substantial evidence based on the entire record.' . . . Moreover, the Sixth Order violated the [Public Utilities] Act" (Illinois Supreme Court 1989, 32–33).

The Court's decision also blocked implementation of the second step of the increase and sent the utility's request for higher rates back to the ICC. The Court did not order the company to refund funds collected pursuant to the Commission's December 1988 order, but it clearly indicated that any amounts subsequently found by the ICC to have been excessive would have to be refunded. As much as $285 million might have to be refunded (Com Ed 1989). Com Ed said it was "very surprised," and Howard Learner of CUB called it "a tremendous victory for consumers." Though nothing was certain at the time, it appeared that the ICC would have to order a $260 million refund for the first step (Grady and Karwath 1989).

Confidence in the Institution of Regulated Natural Monopoly Erodes

The period from 1985 through 1989 had not been a good one for the institution of regulated natural monopoly in northern Illinois. After seriously stagnating for years, Edison's electric sales rose by 13.1 percent (or 3.3 percent per year) over that period (see figure 6.3). Similarly, its operating revenues rose by 15.9 percent (or almost 4 percent per year) and its peak load by a robust 26.6 percent (or 6.7 percent per year). But the most important trend from the shareholders' point of view was one that did not look quite so good: after years of healthy growth, the company's net income and the shareholders' earnings per share of common equity both dropped precipitously from 1988 through 1989 (see figure 6.4). Though the company's net income remained rather large at $737.5 million in 1989, that reflected a $352 million drop from the previous year. With so many court and remand cases underway, and with the City of Chicago exploring its options, it was not clear how well things would turn out for the company or its shareholders.

If Edison was in trouble, so was the Illinois Commerce Commission. Edison's allies complained, of course, that the Commission had not been authorizing the company to earn a fair and reasonable return on its investment. And consumer groups complained that the company's rates were unaffordable. More interesting from the present point of view, however, is that some people began charging that the traditional

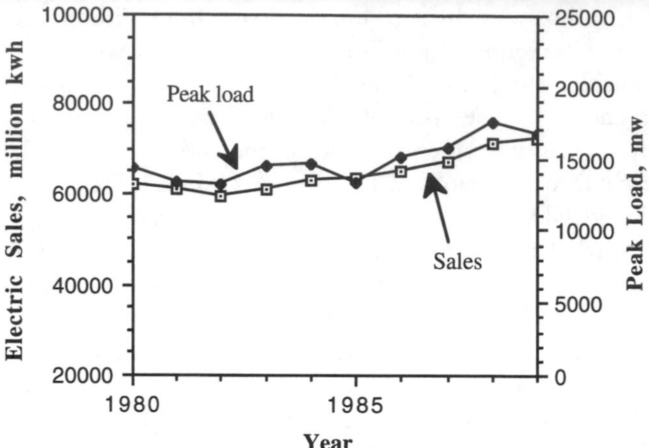

FIG. 6.3 **Edison's Sales and Peak Load Rise Slightly in Later 1980s**

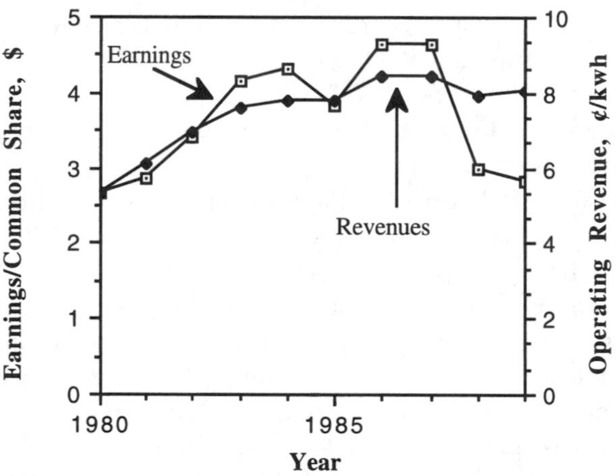

FIG. 6.4 **Edison's "Price" Stabilizes and Earnings Fall in Later 1980s**

approach to rate making was no longer working, that—in effect—the institution of regulated natural monopoly was breaking down. What should be done to respond to those fears, fears that had to do with trust and confidence in the political economy of electric power?

One solution that was favored by some, particularly Edison's allies, was to alter the composition of the Commission itself. In May 1989, the Chicago Association of Commerce and Industry (CACI) charged that recent votes by Commissioners Bushnell and Stone had contributed to a "decline in the quality of public utility regulation in the last few years." Not many weeks later, there were signs that Governor Thompson might not renew Bushnell's and Stone's five-year terms. And a third commissioner, Andrew Barrett, was being considered for a vacancy on the Federal Communications Commission (FCC) (Karwath 1989a). CUB suggested that Bushnell and Stone's predicament pointed out how politics plays a strong behind-the-scenes role in commissioner selection, noting that Bushnell and Stone's chief opponents had been the CACI, Illinois Senate Minority Leader James "Pate" Philip (R), and (according to some) Edison. Edison denied lobbying for the commissioners' removal.

In late October 1989, the governor replaced Bushnell and Stone with Terry Barnich (his legal counsel) and Ellen Craig (his deputy chief of staff). Consumer groups protested, saying that Bushnell's Commission had been the "fairest commission in recent history" and that appointing Barnich simply continued Thompson's tradition of "appointing ambitious young cronies . . . at times when Edison is in need of a rescue." Thompson also appointed Lynn Shishido-Topel, an economist who worked for a Chicago consulting firm, to replace Barrett, who had left to join the FCC (Karwath 1989c).

FIG. 6.5 **New Chairman Terry Barnich**

CUB urged the senate's Executive Appointments Committee to reject the appointments of Barnich and Shishido-Topel on the grounds that neither was qualified to serve on the Commission. Barnich had no utility-related experience, they said, and Shishido-Topel had a conflict of interests because her consulting firm had represented utilities before the ICC. CUB's president, Josh Hoyt, claimed that pressure from utilities kept Thompson from reappointing Bushnell and Stone and that "it looks like Illinois is returning to the bad old days when the utilities called all the shots at the ICC" (Seigel 1989). Consumer objections notwithstanding, the senate confirmed the Governor's appointments and the new commissioners took their seats on the ICC.

By late December 1989, Edison and its allies appeared worried that the company might be facing serious financial difficulty in the near future, and they wanted more fundamental changes made in the legislation that guided the Commerce Commission. A *Chicago Tribune* editorial, for example, argued that the ICC had to restore consumer confidence, end its losing streak in the courts, and tone down the fight between Com Ed and consumer groups. It also advised the state legislature to amend the Public Utilities Act to give the ICC greater flexibility and more options. Consumer groups, on the other hand, saw plenty of evidence that the Commission staff was strongly biased in Edison's favor and that—with the removal of Commissioners Stone and Bushnell—things could only worsen. Com Ed was hoping to get the rate increases behind them so that its management could focus on renegotiating the franchise with Chicago, and consumer and community groups were hoping that the City of Chicago's exploration of options might uncover a way out of the snares of regulated natural monopoly. All eyes turned to Chicago.

Interlude

A Planner's Office in Chicago's City Hall

I climb aboard the Burlington Northern commuter train in Hinsdale, twenty miles west of the Loop. The commuters and I quietly speed toward the Sears Tower, the AMOCO building, the John Hancock building, and the other modernist towers of downtown Chicago. Reading the Chicago Tribune, *I wonder what it must be like to work as a planner in Chicago, long characterized as the epitome of political corruption. A myth? I walk twenty blocks in the rain, scurrying along with thousands of suburban commuters. I dodge the traffic as I pass under the El, and finally arrive at City Hall. The neoclassical design of the exterior reminds me of the Illinois Supreme Court building; but at eleven stories, it is much taller. The interior seems less cared for, though, with broad and busy corridors, old marble floors and walls. I see many African Americans and hear some people speaking Spanish. I walk to the elevator, then ride it past the mayor's fifth-floor office to the Planning Department. Partitions efficiently divide the Department's space into bureaus and offices. I walk to the far-left-corner where the energy planner's office is located. Waiting for the planner to return, I notice stacks of papers and reports piled about. Books line the east wall; state and national maps of electric utility service areas color two other walls. Larger and more private than the other offices, the room has two north-facing windows. I look out, but all I see is another wall of the building thirty feet away and the windows of mayoral staff a few stories below.*

7

Candle, Candle, Burning Brigh[t]
Could Be Chicago Every Nigh[t]

— COMMONWEALTH EDISO[N]

Precinct Captains at the Nuclear Switch?: Exploring Chicago's Electric Power Options

Up to this point, I have focused primarily on planning within the existing institutional structure of regulated natural monopoly. I have argued that Commonwealth Edison's planners and managers were storytellers who authored texts that relied on the "normal discourse" of public utility planning and regulation. Prior to 1985, those texts proved rather persuasive. Through them, Edison persuaded the ICC to embrace the company's far-reaching nuclear expansion plan, to approve a series of expensive rate increases, and to authorize continuation of the company's construction program despite ample evidence that the last few plants were not necessarily needed. Beginning with the Illinois Supreme Court's June 1987 rejection of the Byron 1 rate increase, however, Edison's texts proved less and less compelling. Prodded by the consumer groups' alternate story, the ICC rejected Edison's "negotiated settlement" in July 1987 and—eighteen months later—the Supreme Court overturned the ICC's December 1988 approval of a "settlement agreement."

Note, however, that these texts presumed the existence of a coherent interpretive community. Despite their evident disagreements, the members of this community (Edison, the ICC, consumer groups, and other intervenors) all argued in a context of prior experiences with one another, shared understandings, and a common language. To the extent that Edison's texts failed to persuade, they failed in terms of the agreed-upon conventions of that community.

The ICC's hearing rooms and the Supreme Court's chambers were not, however, the only sites of heated efforts to persuade. Another struggle was taking place in the City of Chicago. Only this struggle did not simply reconstitute the community and culture of regulated natural monopoly. Rather, it consciously considered changing the institution itself, replacing regulated natural monopoly with some new institutional structure. By seeking to change the institution, or to replace it with some new one, this struggle moved away from the normal rhetoric of public utility economics and into the realm of "abnormal discourse." The stories that Edison and others told became more complicated and more passionate.

The interesting issue, which I begin to explore in this chapter and then return to in chapter 9, is whether any participants in this abnormal discourse of increasingly complicated and passionate stories were able to invent a narrative that was capable of evoking a new electric-power institution and persuading the diverse citizenry of Chicago that such an institution was both possible and desirable. Was, in other words, anyone involved in Chicago's electric power planning (including people employed as "planners") able to tell a persuasive story about how electric power could be supplied in a better way, a story that responded to public distrust of both Edison and the City's politicians, or did their narratives all simply reconstitute the existing community and culture of Chicago's local version of electric power politics?

Let me begin answering these questions by briefly recalling the context of Harold Washington's election as mayor of Chicago in 1983.

A New Mayor for an Economically Declining, Politically Divided City

Chicago and its environs were sharply divided in the early 1980s, both racially and economically, and those divisions had been worsening over time. Though the region as a whole was economically vital and dynamic, the city itself had experienced a significant loss of jobs and residents over the past twenty years (Squires et al. 1987). Job losses had been particularly acute on the predominantly black West Side and South Side and in the steel communities of the southeast part of the city. Conversely, the Loop area in downtown Chicago, Dupage County to the west, and the northwestern part of Cook County (in which Chicago is located) had been growing dramatically. While the city was becoming poorer, it was also becoming more black, more Hispanic, and more segregated: in 1960 Chicago was the home of 812,000 African Americans and 2.7 million whites; twenty years later it was the home of 1.2 million African Americans and 1.3 million whites (Kleppner 1985).

These racial and economic divisions had significant political consequences. Prior to 1983 the city had been dominated by a predominantly white Democratic political machine and a coalition (consisting primarily of large corporations and downtown Chicago business interests, including Edison) that stressed the importance of economic growth guided by private investment (Kleppner 1985; Squires et al. 1987). In April, 1983, however, that dominance seemed to have been overthrown. Supported by a mobilized, largely African American, constituency, Harold Washington narrowly defeated former Mayor Jane Byrne and former Mayor Daley's son, Richard M. Daley.

Mayor Washington came to office stressing a progressive reform agenda: he sought to redistribute the city's resources more fairly, emphasize neighborhood economic development, challenge politically powerful businesses, and establish a more open, participatory decision-making process (Moberg 1988). The Mayor quickly discovered, however, that his position was weaker than he would have liked. His opponents, led by Aldermen Edward Vrdolyak and Edward Burke, held a 29–21 majority in the City Council and opposed many of the new Mayor's reform initiatives. Until he could obtain majority backing from the Council, Mayor Washington would find many of his efforts thwarted.

Given Mayor Washington's reform orientation, it comes as no surprise that he focused some attention on the effect that Edison and its rates had on the City's economy. In 1983 Edison, which provides power to the northern one-third of Illinois, was one of the largest investor-owned electric utilities in the country. More important to the residents of Chicago, its power was becoming steadily more expensive. Projecting (and counting on) a steady increase in the demand for electric power, Com Ed had in the early 1970s initiated the nation's largest nuclear power plant construction program. To pay for those plants, Edison had to ask the Illinois Commerce Commission (ICC) to approve a steady stream of rate increases. In 1982, the ICC approved a 17.1 percent increase, the fifth increase in nine years. With five more plants under construction, it looked as though the stream of hefty rate increases would continue for several more years. In 1984, the ICC approved another 6.6 percent increase. Shortly thereafter Edison applied for another 12.2 percent increase for its Byron 1 plant. And there were signs that Com Ed would soon seek a 30–40 percent increase to cover the costs of its last three plants.

Consumer groups strongly opposed each of these rate increases, but without much success. Other consumers, the largest and most mobile, threatened to abandon Edison's system unless the utility lowered their

FIG. 7.1
Mayor Harold Washington Speaking to Supporters

rates. One such customer, at least potentially, was the City of Chicago itself. Edison served the City and its residents pursuant to a franchise agreement that was adopted in 1948 but would expire in December 1990 unless jointly renewed by Edison and the City.[42] Given this agreement, the City apparently had the ability to abandon Edison's system (or at least use the leverage provided by the franchise to negotiate more favorable rates and terms). This leverage was enhanced because City agencies and other Edison customers located within the city limits collectively purchased roughly one-third of Edison's power output at a cost of $2 billion annually. It was in this context that the City started to explore options to remaining on Edison's system.

A Tidal Wave That Threatens to Wash the City's Economic Base out to Sea

On October 10, 1985, Mayor Washington directed his Commission on Energy (which he had created just a few months before) to consider "various mechanisms for the city and other Chicago customers to drop off Com Ed's system and avoid its soaring costs" (Biddle 1985a). Expecting his action to "send a strong signal" that the City was deeply concerned about the utility's rising rates, he told the Commission to look at four options: *a municipal power purchasing authority* (which would buy electricity from another utility at less expensive wholesale

The Mayor suggests that the City does not have to remain dependent on Com Ed. But what does the Mayor's utterance mean? Edison interprets it as a bargaining move and responds with a threat. Community groups understand it to be a call for public ownership.

rates), *a municipal authority* (which would involve buying out part of Edison's system), *wheeling* (which would involve leasing Edison's transmission lines to transmit power to major municipal properties), and *a municipal distribution network* (which would involve buying Edison distribution facilities to transmit power produced by non-Edison power plants).[43] Planning Commissioner Elizabeth Hollander and Charles Williams (Director of Energy Management for the City's Department of Planning) indicated that they expected to sponsor a $100,000–200,000 study of the City's options.

"We own the highway." It is hard to imagine a more arrogant and domineering way of expressing the power that accrues to a regulated natural monopoly over time. Note, however, that Petkus's statement overlooks the fact that Edison's ownership rights are constrained by the franchise.

Com Ed quickly responded by threatening to move its corporate headquarters and $140 million in annual tax revenues out of Chicago. "The city is broke," said Vice President Donald Petkus; "Where do they intend to get the money to buy a new system?" (Biddle 1985a). Edison also indicated that wheeling rights would be an obstacle: "We can't wheel it out, so we don't think we should have to wheel it in," Petkus said; "We own the highway." An Edison spokesman speculated that "these are the opening volleys on the [1990] round of contract negotiations," but Charles Williams said that the City was not looking at the exploration as a "bargaining chip" (Biddle 1985b).

The local news media immediately constructed the issue as a choice between an incompetent city administration and an expensive private utility. The *Chicago Tribune* declared in an editorial on October 16, for example, that there had been a "brownout in City Hall." The administration "has some good ideas for creating competition for Commonwealth Edison in order to cut electricity costs," it said, but "the proposal to form a new power company run by the city is so dimwitted that it casts the others into the shadows" (*Chicago Tribune* 1985).

Note that the City planned to *explore* alternatives. That word implies that the alternatives

These few opening exchanges raised the questions that would shape debate about the issue for the next several years: was the City's exploration of options merely a "bargaining chip" in the expected franchise negotiations, or was the City serious about "dropping off Edison's system?" If the City was serious, did it intend to create a new municipal utility (as indicated by the *Tribune*'s headlines and editorial) or did it truly intend to explore alternatives creatively, an exploration that would take advantage of technological, economic, and regulatory changes that had been sweeping the industry (Kahn 1988; Hyman 1988)? Lastly, how would the City be able to persuade the public and other relevant audiences that the search was legitimate, technically sound, and politically viable, hence not just another instance of "Chicago politics"?

already exist "out there" and that the challenge to City planners is (rather like early explorers looking for the Northwest Passage) simply to find them. What might have happened if the City had asked its planners and interested citizens to help *invent* viable alternatives, to combine the raw materials of the situation in novel ways?

Just two weeks after the Mayor's declaration, the ICC gave the City even greater incentive to abandon Com Ed by granting the utility an 11.0 percent rate increase for the Byron 1 plant. Com Ed predicted that its rates would begin to decline in 1989 and hence that the City had no reason to seriously consider abandoning the utility. Other observers scoffed at that possibility. Howard Learner of the Citizens Utility Board (CUB) said, for example, that "if you really believe Edison's rates are going to go down, you believe in the tooth fairy" (Camper and Schneidman 1985).

The tooth fairy was nowhere to be seen when, in early 1986, Com Ed announced that it might have to apply for a 35 percent rate increase to pay for its last three nuclear plants. That announcement led the Mayor to restate the City's intent to explore alternative sources of electricity and to declare that a 35 percent increase would "deliver a crippling blow to the economic health of Chicago." "Before they hang this albatross around our neck," he said, "I would like to see the commerce commission thoroughly evaluate the chilling effect it will have on the economic health of this city" (Schneidman 1986b).

How "crippling" would the "blow" be? In April the Mayor could not answer that question. Three months later, however, the City's Planning Department released a $15,000 study prepared by Chase Econometrics, which estimated that the Chicago area would lose 85,000 to 112,000 jobs over the next twenty years if the ICC approved Com Ed's requested rate increase. The City's manufacturing sector and its low-income residents would be especially hard hit. Commissioner Hollander indicated that the City's future economic development depended on attracting small businesses with high growth potential and that those firms could not afford higher electricity rates. The rates

Comes the first planning trope: a forecast based on a simple model; that is, on a metaphor.

would also hurt low-income households. Edison quickly criticized the study for double-counting inflation, and claimed that its conclusions were "questionable" and "inaccurate" (Strong and Ziemba 1986). Even so, backed by these data the Mayor once again announced that the City was funding a study of options "so that we can get out from under this untenable situation" (Ziemba 1986). He named Hollander and Robert W. Wilcox, cochairs of his Commission on Energy, to head a new Electric Power Options Task Force.[44]

From the public's point of view, a long period of apparent inactivity followed. The Mayor was fighting a hotly contested reelection campaign, and when he spoke about electric power it was to oppose Edison's requests for rate increases rather than to discuss the more complex issue of exploring options. On March 19, 1987, for example, he told an anti-Edison crowd outside the State of Illinois Building in downtown Chicago that Com Ed's effort to rush consideration of its latest request for a rate increase was "the most outrageous railroading job ever perpetrated" (Galvan and Unger 1987). Washington called for greater citizen input and a thorough exploration of alternatives.

Washington won his campaign for reelection. Almost as important, his supporters won four more seats on the City Council. With the Council now split 25–25, Washington was in a position to cast the deciding vote on major issues. With four more years to look forward to, and with firm backing from the Council, he could now confidently move ahead on more controversial issues.

While the Mayor was running for reelection, his Planning Department staff was quietly moving ahead with its plans to conduct a study of the City's options, and it was working with consumer groups to defeat Edison's proposed rate increase request. In the fall of 1986 it hired the public power–oriented consulting firm of R. W. Beck, which had done similar work for New Orleans and other cities, to conduct the study. Relying in part on Beck's research, the Department joined with several consumer groups to successfully persuade the ICC to reject Edison's proposed "negotiated settlement." But in August, just one month after that defeat, Edison applied for a traditional 27 percent rate increase. The threat of economic harm continued to loom over the city.

The local news media reported nothing further about the City's exploration of options until October, when the *Chicago Tribune* disclosed that "an unreleased report under scrutiny at City Hall," which reportedly cost $75,000, estimated that Chicagoans could save $10 to $18 billion at a cost of $1 to $7 billion over a twenty-year period if the City bought and operated some of Com Ed's facilities (Dold 1987a). Emphasizing that the City's franchise with Edison provided legal authority for the City to acquire those facilities, the Beck study focused on three optional ways of "municipalizing" Com Ed's system:

- purchase all of the utility's assets within the city limits at a cost of $1.3 billion and a savings of $10–12 billion
- purchase 30 percent of the system, excluding certain nuclear plants, at a cost of $6.5 billion and a savings of $18 billion
- purchase 30 percent of the system, including nuclear facilities, at a cost of $6.5 billion and a savings of $15 billion.

"This study was undertaken," the report said, " . . . to explore various options with respect to the furnishing of electricity to the City and the citizens of Chicago," and its purpose "is to summarize the results of the studies undertaken by the Consulting Engineer and to provide information as may be helpful to the policy makers in Chicago" (Beck 1987, I-1). And without providing any details, the report stated that "the costs of electric service [associated with each option] were estimated by using the PWRCOST and APCAP series of computer programs developed by Beck and Associates" (I-14).

The report clearly was not a scientific study, for it did not burden its readers with theory, disciplinary jargon, references, quotations, methodological detail, or other trappings of scientific analysis. It did, however, rely heavily on the conceptual language provided by public utility economics; e.g., "bulk power purchases," "interconnection and coordination services," "discounted present values," "levelized costs of electric service," "reliability" of supplies, "excess capacity," and "disallowances of costs" for nuclear power plants that had been built but not yet placed in the "rate base." It also adopted the antirhetorical stance (the passive voice and a generally somber appearance) that scientific planners prefer. (Bound in gray and 109 pages long, it was printed in black print on white paper, contained seventeen tables and six graphs, and provided thirty-one pages of tables in an appendix.)

By adopting the "antirhetoric" of science—somber in appearance, passive in style, highly quantitative—Beck and Associates appeared to have conducted a scientific study, which in turn lent scientific credibility to the City's claim that establishing a new structure of electric power supply would be in the best interests of the City and its residents. But by limiting its exploration to the well-worn path of public ownership, it exposed the City to charges of trying to reconstitute a culture of political corruption and incompetence.

The Beck report was also clearly not a political analysis, for it did not focus on the political consequences of pursuing alternative courses of action. Deftly avoiding the ticklish question of whether any new public power system might be corrupted by the "Chicago Pols," the report simply said that "In the event the City would decide to pursue any form of the options studied, prompt attention would need to be paid to the development of an appropriate structure for governance, management, and operation of the municipal utility" (I-13).

Neither purely political nor purely scientific, but a hybrid of both, the Beck report exemplified planning as policy analysis: it relied heavily on the language and tools of public utility economics but limited its attention to the City's concerns. Thus the report adopted the City's definition of the problem, compared a few economic consequences of four specific alternatives, and presented a series of contingent recommendations. In the end Beck and Associates concluded that "the potential exists for the City to reduce the aggregate costs of electric service to consumers within the City during the next twenty years" by $1.1 to $18 billion, depending on the option chosen (I-14).

Whether the Beck report exemplified planning as policy analysis or not, it quickly became clear that the City had made a wrong turn in allowing its consultants to focus on such a narrow range of municipalization options. That focus made it much easier to eliminate the municipalization option as a "bargaining chip" in any future franchise negotiations. And losing that chip would put the City in a much weaker bargaining position. Conversely, if the City was serious about wanting to explore (or invent) options, then it learned very little from the Beck study. Rather than inventing options and then probing their economic, political, and technological dimensions, Beck had limited its study to a few variants of one tired old alternative. Last, by authorizing Beck and Associates to focus on municipalization, the City opened itself to the interpretation that Edison favored: that the City was trying to expand its patronage system and continue the game of Chicago politics.

Just a few days after disclosing the Beck report, for example, the *Chicago Tribune* editorialized (October 30) that the City should negotiate the best possible deal with Edison, but that it should not seriously consider establishing a municipal utility. The editorial argued that the risks of "unplugging" Com Ed would outweigh any potential benefits for several reasons, including the risk that "bureaucrats and political pressures" would largely determine the municipal utility's spending and rate structure. In sum, the *Tribune* claimed that "the thought of this city administration, or any future one, managing a complex electric system that delivers the power critical to daily commerce and living is mind-boggling."

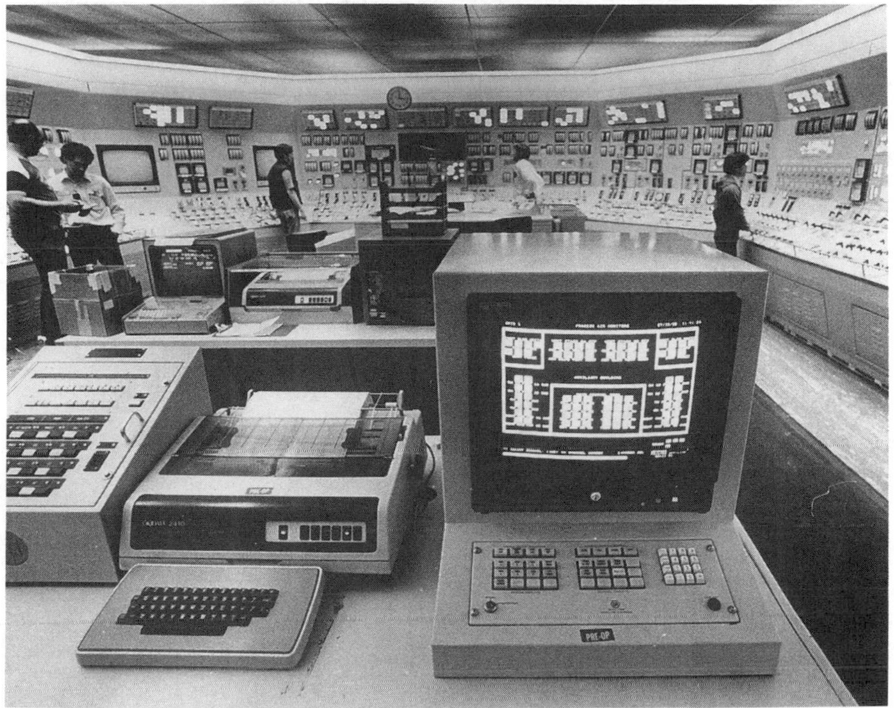

FIG. 7.2 **Control Room at One of Commonwealth Edison's Chicago Generating Stations**

When the Mayor formally released the Beck Study at a press conference on November 4, he and his staff seemed aware that they had created difficulties for themselves. "Six rate hikes in 10 years and a proposed 27 percent rate hike represent a tidal wave that threatens to wash the city's economic base out to sea," Mayor Washington declared (Dold 1987b). But, frustrated by accusations that the City was just trying to add eight thousand more jobs to the patronage system, he also stressed that additional studies were needed before the City could decide whether to purchase Edison's facilities. "We've been disappointed that there has been a rush to judgment about our motives," said Commissioner Hollander; "We think it would have been irresponsible not to look at our options" (Dold 1987c).

The Planning Department did more than express frustration. At the Mayor's press conference, it also distributed an overview of the Beck report. The cover page of that overview (see fig. 7.3) symbolically appealed to the anti-Edison coalition for support and selectively portrayed the Department's view of the City's relationship to the utility: low on

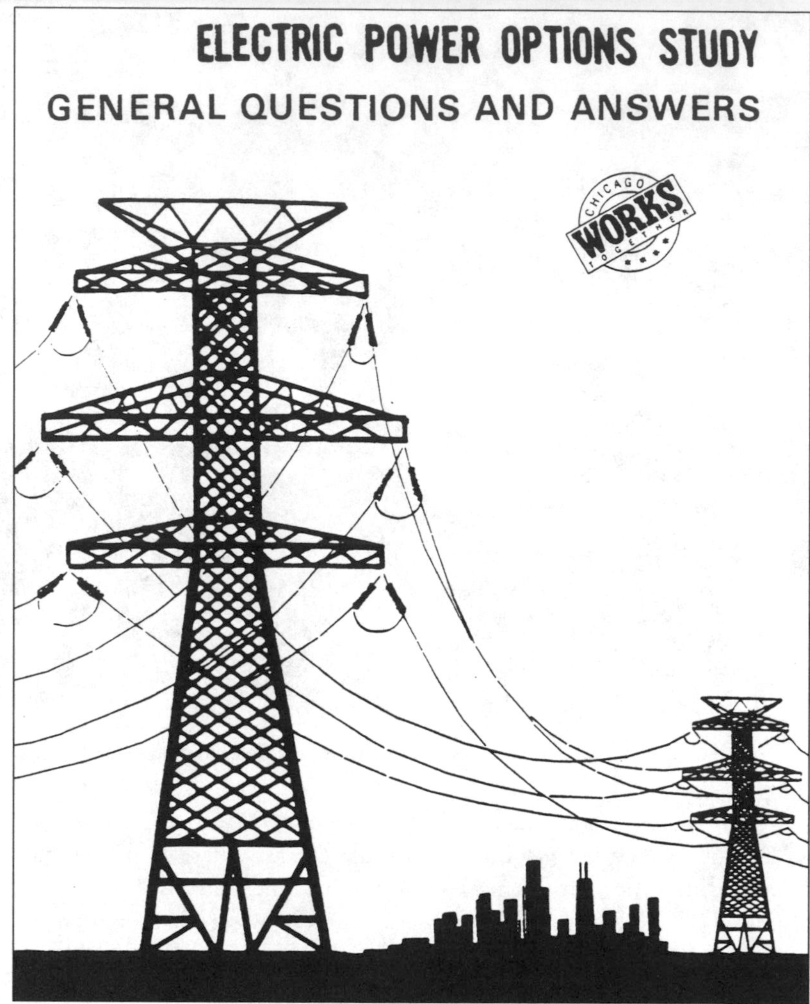

ELECTRIC POWER OPTIONS STUDY
GENERAL QUESTIONS AND ANSWERS
CHICAGO
WORKS
TOGETHER

FIG. 7.3 **The Chicago Department of Planning's Overview of Electric Power Options**

the horizon lay a small representation of downtown Chicago's skyline, and looming above that skyline were electric transmission lines that stretched between two immense transmission towers. The cover clearly implied that Edison has power over Chicago, but the title of that cover page boldly proclaimed that Chicago is free to explore alternatives to continued domination by the utility. The remainder of the overview reinforced the rhetoric of the cover page. In it, the Department argued that Edison mismanagement had caused electric power rates to soar, that those soaring rates had caused enterprises to avoid investing in

Chicago, that Com Ed's "traditional monopoly supply arrangements" had "stifled" competition, and that the City had the responsibility to explore its alternatives. Moreover, the overview also portrayed the Department as a publicly accountable, politically responsive organization that seeks technical advice from experts, solicits advice from a broad range of interests in Chicago, is sensitive to the political needs of the Mayor and the aldermen, and seeks to keep the public informed about what the Department is doing.

Stop the Bureaucrats and Activists Now!

Commonwealth Edison responded to the Beck report with technical critiques, political manipulations, and advertisements in the news media. The political response actually preceded the Mayor's formal announcement by several weeks. At the same time as it disclosed the Beck study, the *Tribune* also reported that the U.S. House Ways and Means Committee had just approved legislation that would keep the City from using tax-exempt bonds to fund the purchase of Com Ed equipment. Introduced by Chairman Daniel Rostenkowski, a Democratic congressman from Chicago, this legislation would reportedly increase the cost to the City of acquiring Edison's facilities by as much as 50 percent.

Edison's response to the Beck report signals quite clearly that the public discourse about Chicago's exploration of options will be simultaneously technical and political. That it is normative as well soon becomes clear. One begins to suspect that the rhetoric of this abnormal discourse will simply reconstitute the culture of Chicago politics.

The Mayor's supporters quickly accused Edison of pulling strings. Rostenkowski, they charged, had introduced this legislation at the request of James J. O'Connor, Com Ed's chair and Rostenkowski's long-time friend (Moberg 1989). "Our own congressman is shooting his own city in the foot," said Alderman Raymond Figueroa. "Rostenkowski has led the charge to protect the monopoly stranglehold Commonwealth Edison has over Chicago's electricity consumers" (Dold 1987d). They demanded that Rostenkowski withdraw the legislation on the House floor.

Rostenkowski denied the charge of course. "I'm not in bed with any utility," he said; "I'm trying to stop the revenue drain so we're doing something about this deficit" (Dold 1987e). Rostenkowski said that he had not been lobbied by utility representatives, that the measure was drafted to reduce the federal deficit, and that he might delete the utility provision if he could be convinced that Beck's savings projections were legitimate.

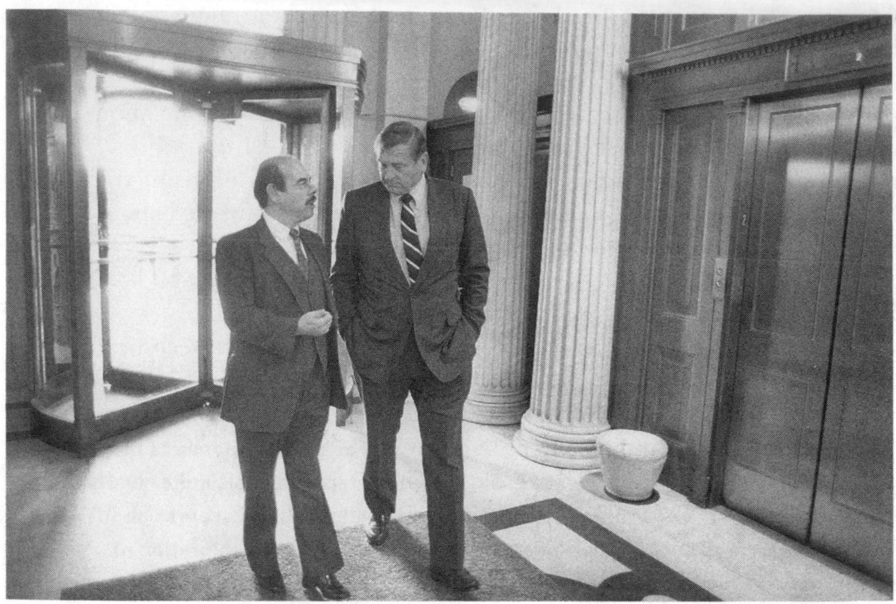

FIG. 7.4 **Representative Dan Rostenkowski, right, on his Way to the House Floor**

The City's supporters tried to persuade Congress not to retain Rostenkowski's provision (Galvan 1987; Dold 1987f). Ultimately, however, they failed. On December 18, House and Senate conferees voted to retain the Rostenkowski provision. Afterwards, Democratic Congressman Charles Hayes sharply criticized Rostenkowski: "Chicago is getting tattooed all the way around," he said, noting that the conference report had included exemptions for a proposed utility buyout on Long Island and for expansion of public utilities that are more than ten years old. "If Danny could find a way to build in protection around New York and Texas," Hayes said, "he could have done it for Chicago. He just didn't really try" (Dold 1987g).

Edison's second response to the Beck gaffe was technical. An hour after the Mayor formally released the Beck study, Edison officials held their own press conference at which their consultant, Harold Axelrod of Planmetrics, condemned the study for being "riddled with faulty thinking and highly improbable assumptions" (Dold 1987b). City officials had claimed that fifteen midwestern utilities were interested in supplying power to the city, and that cogeneration and energy conservation would make up any potential supply shortfall after the mid-1990s, but Axelrod said that the City was "being led down the garden path." Edison officials also emphasized that suburbanites could end up "subsidizing" any City "takeover" (Dold 1987c). A few days later Edison told reporters that the Beck Study had drastically underestimated the cost of its power grid and generators in Chicago and that those facilities would cost roughly $1.5 billion more than Beck had estimated (Arndt 1987d).

Edison did not publicly release Axelrod's report. In late January, however, it did present the results of that study at a meeting of the Western Society of Engineers (Dold 1988b). Countering the Beck study, Planmetrics reported that only nine of forty midwestern utilities surveyed would be likely to supply power to Chicago if the City owned the power grid, but that those utilities could not meet all of the city's needs past 1995. It also reported that rates in Chicago would be 23 percent higher under a city power authority than they would be if the ICC had approved the utility's negotiated settlement. Last, Axelrod's report indicated that existing transmission capacity might not be adequate and that the Beck study had failed to count $10–12 billion additional construction and rehabilitation costs.

Edison's third response was to conduct an expensive and aggressive advertising campaign. For at least eighteen months following the Mayor's press conference, Com Ed advertised its case—on radio, television, newspapers, and through a series of neighborhood meetings—against the City's effort to "take over" Com Ed's facilities.[45] In these advertisements, Com Ed consistently portrayed itself as a model citizen of a world-class metropolis whose growth and progress vitally depended on a steady supply of safe, clean, reliable, and reasonably priced electric power. It further portrayed itself as an experienced, skillfully managed corporation that was trying to protect and advance the interests of its consumers and shareholders and of the Chicago region as a whole. At the same time Com Ed also tried to de-legitimize any challenges to its expertise. Regardless of the City Council's legal authority and responsibility to reconsider the franchise with Com Ed in the very near future, and disregarding the evident right of the City's citizens to voice their opinions about what the Council should do, Com Ed disparaged the City's planning staff and many of its citizens.

What do the words "take over" mean? Do they not imply that the City—with its presumably faceless bureaucrats and corrupt aldermen—was on the verge of seizing something that belonged to someone else, that it intended to infringe on Edison's private property rights?

Let me present three striking examples of this advertising campaign. Here is the "Taxi" radio spot that Com Ed ran in late 1987 and in 1988:

Here Edison speaks through the voice of an "average cab driver" to an implied listener who knows more about local sports than about Com Ed. The implied listener is persuaded rather easily despite the likelihood that most actual listeners do not like the rate at which Edison's prices have been increasing.

Guy:

Ya know, I can't rightly explain it, but nobody seems to know more about sports than your average cab driver. So I wasn't surprised the other day when mine turns to me and asks what I thought about all this talk over changing managers. I said, well, to tell you the truth, I thought pitching was the biggest problem. Oh, no he tells me. He meant how people have been talking about the city managing its own electric company. I said, I thought Commonwealth Edison already did that. Oh, they do, he said. Been doing it for over a hundred years, in fact. Why this summer alone they set seven all-time peak load records and they did it without those power shortages like the East Coast had. I said well then why do some folks think the city can do a better job? . . . He said, oh most folks don't. In fact, in one survey, 2 out of 3 said they were against it. I said, well I'm no expert, but seein' as how the people have already spoken, maybe the problem is that some folks just aren't listenin'.

ACR:

Commonwealth Edison. We're there when you need us.

Com Ed also ran a very interesting "Windmill/New" television spot which attempted to persuade the public that the City's "takeover" campaign would cause Chicago to abandon an era of progress and prosperity (fueled by Com Ed's power) and to enter an era of risk, decay, and declining expectations (all because the City was going to force people to use untrustworthy technologies). The original ad contained twelve separate scenes. The first displayed the beauty of downtown Chicago at night, illuminated by electric lights. With music in the background, an off-screen narrator said: "Every hour of every day . . . Commonwealth Edison works to keep down . . . the cost of the electricity we produce . . . We'll even go outside Illinois for energy . . . If we can get it for you at a better price." Then the scene suddenly shifted to the upper three stories of an older apartment building. On the roof was a rather old looking windmill. The lights in the three rooms directly below the

Com Ed uses the image of decaying inner-city apartments to convey what would happen if the City took over the power system. Is there not a bit of unintentional irony here? The decay is not something that *will* happen. It has already happened . . . during the time that Com Ed managed the electric power system.

LEO BURNETT COMPANY, INC.
AS FILMED AND RECORDED
(9/95) "Windmill/ New/ Rev." :30
COMMONWEALTH EDISON CO.
YCHI2843

1. (MUSIC: UNDER THROUGHOUT)

2. (AVO): Every hour of every day Commonwealth Edison does everything possible...

3. to make sure the eight million people in northern Illinois...

4. have all the electricity they need...

5. no matter...

6. how great the demand.

7. In fact a more reliable source of electricity...

8. would be hard to come by.

Commonwealth Edison
We're There When You Need Us

9. Even if you made it yourself.

FIG. 7.5 **Commonwealth Edison's "Windmill/New" Television Advertisement**

windmill were off in the first scene, then on, then off, then on. Again off-screen, the narrator said: "The fact is, more reliable . . . more economical . . . electricity . . . would be hard to come by . . . even if you made it yourself." Superimposed over the windmill in the last scene were the words "Commonwealth Edison—We're There When You Need Us." Figure 7.5 shows a later version of that advertisement.

"There when you need us," Com Ed also began distributing to its consumers a four-page flyer that warned of impending doom in the event of a City takeover (Com Ed 1988). The first page tells the whole story. Printed in yellow letters (caution), trimmed with a

Here Edison speaks by synecdoche: substituting "bureaucrats and activists" for the whole range of citizens who might want to influence the course of the City's exploration. By separating

the city government from its citizens, the rhetoric once again reconstitutes the culture of Chicago politics. Notice, as well, that the flyer implies that consumers have greater control over Edison's system than citizens do over Chicago government. Perhaps, but earlier chapters of this book lead me to be skeptical about that claim.

thin red line (stop, danger), on a black background (darkness, no light), are the words "Candle, Candle, Burning Bright, Could be Chicago Every Night . . ." Claiming that "a few bureaucrats and activists" are trying to have the City "take over" Edison's system, the flyer charged that such a takeover was a "seriously flawed bad idea" that would result in an inefficient, politicized electric power system that would dole out favors "at the whim of the City Council." [46] It is time to "STOP THE BUREAUCRATS AND ACTIVISTS NOW!" the flyer said, and let political representatives "know about your opposition to a City takeover of your electric system."

The Death of the Mayor and a Cautious Renewal of the City's Exploration

By mid-November 1987, it appeared as though the Beck report had done more to obstruct the City's exploration than to expose a variety of possible paths for it to follow. Even so, there were good reasons to believe that the City could rebound and regain the initiative. Mayor Washington had already asked the City Council to include $500,000 in the 1988 budget for continuation of the City's exploration. About one-half would be used to develop a plan about how to create and operate a municipal authority, while the other half would be used to conduct in-depth studies of Chicago's future electricity needs. Charles Williams stressed that municipalization was

> clearly feasible, financially, legally, and logistically. . . . But we can't dismiss the fact that it's a major undertaking. We have not made a decision whether this is a direction we want to go in. These are questions we absolutely have to answer before we make the final determination. It indicates the city is very, very serious. The administration has made a commitment to exploring all the possible options that the city has to reducing the price of power. (Dold 1987d)

The likelihood that the City would regain the initiative was thrown into doubt, however, by the sudden and completely unexpected death of Mayor Washington on November 25. Overcome with grief, many of the city's residents mourned his death. Mourning in their own way, the City's aldermen struggled to appoint a temporary replacement. In a chaotic session that revealed deep divisions among Washington's supporters, the City Council elected Alderman Eugene Sawyer as acting mayor.

FIG. 7.6
New Mayor Sawyer at a City Council Meeting

Mayor Washington's death had a dramatic effect on the City's exploration of options. As time passed it became clear that the acting Mayor, who would soon have to campaign in a special mayoral election and who lacked Washington's firm base of political support, would act much more cautiously than Washington. But that caution was not immediately evident. Sawyer and the City Council kept money in the 1988 budget for the City's exploration, in spite of firm opposition from Edison and its supporters.

Edison's opposition took several forms, one of which was another public relations gambit. In mid-December, just two days before the City Council was to vote on whether to appropriate funds for additional studies, the Chicago Association of Commerce and Industry (CACI) released the results of a survey of Chicago businesses conducted by McKeon

Here Edison's supporters speak by synecdoche: they produce a survey that purports to reflect the views of Chicago

businesspeople. Note how the survey creates a community in which businesses and "the City" are separate entities. Chapter 8 probes the tropal nature of this survey in considerable detail.

and Associates. According to Samuel Mitchell, president of CACI, the survey showed that nearly 30 percent of the 454 firms surveyed would relocate outside of Chicago if the City became the sole provider of electricity and thereby cost the city 250,000 jobs. "We think it would be prudent for Mayor Sawyer and the City Council to think long and hard about whether it is worth spending an additional half-million dollars to study a proposal which sends a negative signal to the business community," Mitchell said. Consumer groups condemned the study as "sheer propaganda" and urged the City Council to keep the money in the budget (Recktenwald 1987).

A second form of opposition was to suggest a different way to fund the study. On the day before the City Council was to vote, Edison Chairman O'Connor offered to contribute $250,000 to the study, subject to the condition that it be conducted by one of the Big Eight accounting firms located in Chicago (Dold 1987g). Presumably such a study would be more "objective" than one conducted by a firm selected by the City's Planning Department.

A third was to mobilize Edison's aldermanic supporters. Edward Burke, one of Washington's most outspoken opponents, took the lead. At one point he joked that "[t]he bozos they have in the Revenue Department can't collect unpaid parking fines or water bills now. If these characters get a hold of a nuclear reactor, Chernobyl will look like a walk in the park" (Dold 1987g). It would be better, Burke said, to let Com Ed fund the study with "no strings attached" and to use the City's money "to feed the hungry or house the homeless, or . . . to increase energy relief payments for the poor" (Devall 1987a).

None of these efforts succeeded, however. Proponents of the exploration argued that it was needed to improve the City's bargaining position in the forthcoming franchise negotiations. Alderman Bobby Rush said, for example, that the study would allow the City to do some "comparison shopping" (Devall 1987b). Supporters outside the Board also stressed the importance of the study for future bargaining: Robert Creamer of the Illinois Public Action Council (IPAC) emphasized that "if the city does not develop alternatives . . . Edison will have all the chips," and Susan Stewart of CUB said that "cutting $100,000 now, when you could potentially save millions of dollars for your constituents is ridiculous" (Devall 1987a). Their arguments carried the day. On December 29, the Council voted 28–11 to keep $400,000 in the budget. The exploration continued.

On the same day that the Council approved the budget, Mayor Sawyer declined Edison's offer to help fund the study, stating that "because of the very sensitive nature of the proposed study, the wide range of direct and indirect interest in its outcome, and the special relationship of your company to the issue, I feel it would be inappropriate for the Edison Co. to provide the study funds." A spokesman for Com Ed said that Sawyer's action meant that "we will undoubtedly have two studies done. The city seems determined to get a study that will produce answers it wants. We feel duty-bound to have our own study conducted by a more objective firm" (Devall 1987b).

On January 9, 1988, a *Tribune* editorial praised the City for voting to examine its options and not accept Com Ed funding, but it also condemned the idea of a City-owned electric power system. A month later it ran three lengthy review articles about "pulling the plug on Edison" (Dold and Arndt 1988a, 1988b, 1988c). These generally enlightening articles argued that the City would have to overcome a formidable array of obstacles before it could set up a "City Hall Electric Co.," but that a study of options would at least enhance the possibility of gaining concessions from the utility.

After these three articles appeared, the local news media essentially stopped reporting on the City's exploration of options. The *Tribune*, for example, ran only nine short articles and one editorial over the next fourteen months. To ordinary people, it must have appeared that virtually nothing was happening. All they learned was what they heard and saw in Com Ed's ads.

It was not true that nothing was happening. The Mayor's weak electoral position induced caution, but his Planning Department slowly began the new study of options. In June 1988, it sponsored a "Supply Options Workshop" that solicited advice from energy experts from around the country, including Ralph Cavanagh of the Natural Resources Defense Council (NRDC) and Charles Komanoff of Komanoff Energy Associates (KEA). Though not reported in the *Tribune*, this workshop apparently affected the Department's view significantly. Charles Williams later said that Cavanagh had presented "a very articulate vision" of what conservation means for utilities, their customers, and national energy strategy. Impressed by Cavanagh's radiant self-confidence and command of facts and figures that "can almost be intimidating," Williams also said that Cavanagh had "articulated a tool" that no one else had thought of: a compact; that is, a short-term franchise that made future renewal conditional upon Edison's performance (Williams 1990).

While the City was quietly moving ahead in private, in public it was defending itself against Edison's very successful effort to frighten people with the specter of precinct captains running nuclear power plants. In mid-October, for example, Mayor Sawyer explicitly rejected any possibility that the City might buy and operate Com Ed's nuclear facilities. "Our goal is lower, long-term electric costs and greater competitiveness in the energy markets," Williams said. "Not even in our wildest dreams do we think about running nuclear plants" (Dold 1988c). But the Mayor kept alive the option of buying Com Ed's transmission and coal-fired generating systems in the City. The City would turn to that option as a "last resort," he said, if Edison failed to agree to major concessions in upcoming talks on renewing its franchise (which the City expected to begin by the fall of 1989). To help prepare for those talks, the Mayor included $150,000 for final research on energy options in his proposed 1989 budget. Further, the Planning Department was negotiating a contract with a public relations firm that would explain the City's position to the public. Stung by the success of Edison's advertising, Williams emphasized that the City needed "to educate the Chicago public about the choices we're looking at, and let the public provide input on the services they want" (Dold 1988c).

Looking at "choices" would be a futile exercise, however, unless the City could find a way to circumvent the obstacle Congressman Rostenkowski had created. Help came in the fall when the privately sponsored Chicago Energy Commission (CEC) suggested a creative (but complicated) purchasing scheme. The City would buy a portion of Com Ed's system at the cost stipulated in the franchise, then immediately resell it at a higher, market value to a private investor group that would buy it with about 20 percent equity investment and 80 percent debt. The buyer would then lease back the system to a special-purpose public corporation. The lessor-owners would be able to take accelerated tax depreciation and thereby partly get around the Rostenkowski obstacle. That public corporation would then hire an experienced private management firm, employing technicians previously employed by Com Ed, under a contract that specified expected performance (Mayor's Task Force on Energy 1989a).

Rather than accepting the artificial duality between public and private, the CEC consciously invents an alternative that blurs that distinction. At this point in the narrative, however, it is not yet clear whether advocates of the plan will be able to talk about it in a manner that Chicago citizens find persuasive.

Was such a scheme feasible? How would it compare with the City's other options? To what extent would energy conservation help? To help it answer those questions, the City hired a team headed by Komanoff Energy Associates (KEA) to conduct a series of studies.[47]

The team did not begin its work until early 1989. In the meantime, Edison continued to advertise heavily against the City's "takeover" effort. It also sponsored another McKeon and Associates poll which reported that two out of three registered voters in Chicago opposed a City takeover. The *Tribune* characterized Edison's poll as a "preemptive strike" against referenda that were scheduled to be on the November 8 ballot. A spokesman for Edison argued that the language of the referenda predetermined the vote, hence that the utility would be wasting its time trying to oppose them. "This thing is sort of a Trojan horse," said an Edison spokesman. "The way it's worded, they would come up with a majority anyway and say we pulled out all the stops to beat it" (Dold 1988d).

The referenda were the result of efforts by the Chicago Electric Options Campaign (CEOC), a coalition of twelve organizations that had been formed in late 1987.[48] Labeled "extremists" by Edison, CEOC members denounced the utility and pressed the City to continue the exploration of options. In a variety of newsletters, public reports, and public demonstrations, these groups consistently portrayed themselves as nonprofit, nonpartisan, citizen-based organizations who wanted to "educate" the public about Edison's mismanagement and about the need to achieve fundamental changes in how electric power and other services are produced and delivered in the Chicago area (CEOC 1989; Fremon 1988; Hellwig 1989; Kraft 1988).

Countering Edison's earlier surveys, the CEOC sponsors referenda. Note the wording, however. It's not so much that the language biases voter responses, as it is that the language shapes attention. When the public is confronted with difficult technical issues, do not all surveys and referenda shape perceptions more than measure them?

The CEOC's November 8 referenda asked, "Should the City of Chicago seek lower electric rates by taking actions such as purchasing power from the lowest-priced sources available, acquiring power sources in the City, and/or joining with other communities to initiate a Northern Illinois Power Authority?" Josh Hoyt, leader of the referendum campaign, said that "we chose this tactic because politicians only understand two things . . . Dollars and votes. Votes are their meal tickets. Without the vote, they get no dollars. When you go door to door, you talk their language" (Burleigh 1989). Placed on the ballot in fifteen wards, the referendum was approved by a majority of roughly 110,000 to 18,000 (Hellwig 1989).

Octopus, scaremonger, peddlers of panic, bureaucrats and activists, takeover, tidal wave, Trojan horse, economic health of the city. By now it should be clear that planning rhetoric is, at least in this case, deeply figurative. But it is not just the "political" imagery that turns conversation one way or another; models, forecasts, surveys, and other figures of argument turn them as well.

The degree of support registered for the City's exploration in this referendum suggested that the City's relationship with Edison might be a key issue in the mayoral election that was scheduled for the spring. Com Ed was disseminating erroneous and misleading information, they argued, and it was important to ensure that candidates for mayor would agree to continue the City's exploration. In one of its newsletters, the Nuclear Energy Information Service (NEIS) claimed that Edison's aggressive $630 thousand advertising campaign (at the ratepayers' expense) constituted "deliberate deceptions designed to mislead the public" (Kraft 1988). In a report to consumers distributed in November 1988, David Fremon of the Chicago 1992 Committee (which was closely allied with the CEOC) charged that Com Ed was "a utility octopus whose main concern is providing its stockholders maximum profits" and that "Com Ed scaremongers" and "peddlers of panic" are trying to maintain Com Ed's "stranglehold" and "smokescreen the other options available to the city short of complete takeover." Fremon further charged that Representative Rostenkowski was in Com Ed's hip pocket: "How else," Fremon wrote, "can one explain his sponsorship of a bill whose sole effect was to cripple the city in its negotiations with Commonwealth Edison?"

By the winter of 1988, the campaign for electing a permanent replacement for Mayor Washington was in full swing. At least five candidates initially ran for the office: acting Mayor Sawyer, Cook County State's Attorney Richard Daley, and Aldermen Danny Davis, Timothy Evans, and Larry Bloom. But, Fremon and the referenda notwithstanding, none of them made the franchise a major issue (Moberg 1989). Only occasionally did any of the candidates speak about it. In early December, Alderman Davis called on Com Ed to open its books for city inspection and criticized the utility for labeling its opponents "extremists and activists" (Kaplan 1988). (A spokesman for Edison responded that Com Ed was not against discussing the future of the franchise, but that it was against people who wanted to "get rid of us.") In late December,

Candidate Daley promises to "fully assess all the options." Chapter 1's critique of rational planning provides several good reasons to be skeptical of his promise (whether he meant it or not). He will incompletely assess a narrow range of alternatives. Which alternatives? According to what criteria?

Alderman Larry Bloom declared that Cook County State's Attorney Richard Daley had exhibited "poor judgment" and "sold out" to Com Ed in December 1986, when he agreed to sign the negotiated settlement (see chapter 5). "It could be the most important economic development decision that will be made by the next mayor," he said; unfortunately "Richie Daley has shown that he is willing to cave in to Commonwealth Edison, that he's willing to negotiate away Chicago's economic future" (Hardy and Dold 1988). "Bah, humbug," said a Daley spokesman. "During eight years in office, Richard Daley has fought virtually every traditional rate increase sought by Edison and has participated in cases that have helped save consumers over $500 million" (Hardy and Dold 1988). Daley himself pledged to be a "tough negotiator" who would consider all options "except using city employees to run the utility." Daley also promised to give notice to Com Ed of the City's intention to terminate the franchise and to acquire the facilities, and to "provide the necessary financial and staff resources to fully assess all of the options." He promised open public review and participation and a policy of "four Cs"—competition, conservation, cogeneration, and cooperation between the City and its supplier on a "least-cost energy plan" (Moberg 1989).[49]

A major reason for this lack of attention was that the 35 percent rate increase originally feared back in 1985 never materialized. True, Edison had applied for a 27 percent increase in mid-1987. But—as shown in chapters 5 and 6—that case dragged on for months and ultimately resulted in a much smaller increase. In December 1988 the ICC authorized Edison to increase its rates by approximately 8 percent in two steps. Rather than being engulfed in a "tidal wave" that would "wash the city's economic base out to sea," the city's residents were nudged by a relatively gentle flow.

Even so, Edison's rates were still quite high relative to others in the Midwest, and the City still had good reasons to explore its options. Politically insecure, however, Mayor Sawyer needed to proceed cautiously. Apparently sensitive to claims that Washington's Electric Power Options Task Force was too radical, he created a new Energy Task Force in late January 1989. Proponents of the City's exploration worried that this action signaled that Sawyer might not be willing to "play hardball" with Com Ed in the renegotiations: "Clearly, Mayor Sawyer would like a task force that is more amenable to Commonwealth Edison," said CUB president Josh Hoyt. Cautious though he was, the Mayor still asked the City Council to approve a $100,000 public information contract with Calmar Communications to counter Edison's advertising campaign (Dold 1989a).[50]

In February, Sawyer appointed thirty-eight members to his new Energy Task Force. Chaired by Robert Wilcox of the private Chicago Energy Commission, the Task Force quickly formed itself into two subgroups (Technical Research and Public Outreach) and scheduled a series of public meetings with Edison, consumer groups, experts, and others. It planned to conduct public hearings in October and then deliver a final report to the Mayor and the City Council in November. Just as the Task Force was beginning to meet, however, and just when the City's consultants were beginning to submit the first of their technical reports to the Planning Department, another event took place that would have a dramatic effect on the City's exploration. In April 1989, the people of Chicago chose a new mayor.

The continuing effort to appoint a "representative" task force might alert us to the tropal nature of such entities. It is not so much that such groups actually "represent" the public as it is that they form part of a larger effort to persuade the public that actions about to be taken are legitimate.

The Abnormal Discourse of Institutional Change

From April 1985 through April 1989, the government of Chicago, some of its citizens, and Commonwealth Edison engaged in a heated public argument about electric power planning. Unwilling to blindly accept the institution and rhetoric of regulated natural monopoly, and prodded by community and consumer groups, Mayor Washington and some of his staff consciously chose to explore alternatives to continuing on Commonwealth Edison's system. Before returning to the exploration in chapter 9, I want to linger here for a moment and to reflect upon the nature and conduct of that exploration during the Washington and Sawyer years. Here I want to consider whether any of the participants in this discourse were able to imagine and articulate a new path into the future and whether any of them were able to persuade the diverse citizenry of Chicago that such a path was both possible and desirable. Or did they all rely on old rhetoric and old stories that simply reconstituted the community and culture of Chicago politics?

A new path. To take such a path implies turning off the well-worn trail of the past. The trail of the past can be thought of as the flow of utterances and replies (or complicated story) leading to a particular point in time and space. So, let me quickly reconstruct the flow of utterances and replies in Chicago's exploration of electric-power options from 1983 through 1989. Then it will be possible to consider how technical reports, political pressures, and unique events shaped the direction of that flow.

The flow began with the election of Mayor Washington and with Edison's request for rate hikes to cover the cost of its last four nuclear power plants. Objecting to Edison's proposed hikes, the Mayor sponsored two studies (by Chase Econometrics and by R. W. Beck) which indicated that the hikes would cost the city thousands of jobs and that those job losses could be avoided (at a net benefit to the City) by municipalizing Edison's system in Chicago. Edison replied to the Beck study with a threat to move its headquarters from the city, with a consultant report of its own, with expensive advertising in the news media, with surveys sponsored by the CACI, and with (though this claim is contestable) legislative manipulations by Congressman Rostenkowski. The net effect of these replies was (1) to ridicule and de-legitimize the City's exploration by implying that it would place precinct captains in charge of nuclear power plants, and (2) to dramatically increase the cost of municipalizing the system. Offended by Edison's replies, and knocked off balance by the unexpected death of Mayor Washington, the CEOC then sponsored referenda that sought to restore legitimacy to the exploration and to put pressure on mayoral candidates to support it. The City's planners sought to restore their political legitimacy by creating a new Energy Task Force and to restore their technical credibility by hiring a persuasive and well-respected technical team to explore the City's alternatives more thoroughly. The new mayor, Richard Daley, said little about the exploration during his campaign, and few if any people knew what he intended to do.

This flow of utterances and replies can, on one level, be characterized as a move away from the normal discourse of public utility economics, into an abnormal discourse that expressed the community and culture of Chicago politics, and then—perhaps surprisingly—into an abnormal discourse that sought to reconstitute that community and culture (at least with regard to electric power) by inventing a new institution and accompanying rhetoric.

The Beck report walked the well-worn path of public utility economics, a path in which the landmarks and direction are clear. By adopting a somber and passive style, and by relying heavily on the rhetoric of public utility economics, Beck and Associates conveyed the impression of having conducted an objective, scientific study. This "antirhetorical" stance supported the City's effort to persuade elected officials and the public at large that establishment of a new electric power structure would be in the best interests of the City and its residents. The report sought to make its readers think that the issue was technical, a simple matter of monetizable costs and benefits.

That the issue was not simply technical and that it would not be argued in the normal discourse of public utility economics quickly became clear. Setting aside the policy analytic rhetoric and conventions of the Beck report, Edison responded with the rhetorical conventions provided by the normal discourses of politics and advocacy. Suddenly the City's explorers entered the dense foliage, the heart of darkness, that is Chicago politics. To outsiders, the jungle seemed unfathomably complex, but insiders had no difficulty following the trail. Edison, community groups, and the City's officials all spoke the technical-normative-political rhetoric of the place.

Condemning its opponents as "bureaucrats and activists" who wanted to "take over your electric power system" and throw the Chicago region into an era of economic decay, and relying heavily on provocative symbolism (the windmill on the dilapidated apartment), Edison sought to prefigure the opinions of vast numbers of voters and consumers with a barrage of advertising, and it sought to drastically alter the feasibility of municipalization by (opponents charge) persuading Congressman Rostenkowski to promote legislation favorable to the utility. (Of course, it was not quite that simple. Edison also responded with a hired consultant who said that his data showed that the City was "being led down the garden path" and with surveys that showed that the city would lose 250,000 jobs if the City took over Edison's system.)

Edison's advertising and political barrage led Edison's opponents to respond, partly as advocates and partly as political entrepreneurs. They tried to sway the hearts of those same voters and consumers with claims that Edison was engulfing them with a "tidal wave" of rate hikes and that Edison was a "utility octopus" whose "scaremongers" and "peddlers of panic" were keeping the people of Chicago in the utility's "stranglehold." They also acted as political entrepreneurs, trying through the CEOC's referenda, workshops, presentations to neighborhood groups, and related efforts to mobilize political support for the City's exploration of options.

Whew! One cannot help but be impressed by the furious passion of Chicago-style political-economic rhetoric. One cannot help but sense that Edison and its opponents became frozen in a passionate embrace marked by time-honored and easily predictable gestures, grunts, whispers, and blows. Their rhetoric, in other words, largely continued the community and culture of Chicago politics.

This disappearance of modernist, "scientific" planning and its replacement by the abnormal discourse of passion, reason, and power could easily be interpreted as just another retelling of the old story about how "planning" always loses out to "politics." According to this view, technical analysis on the basis of public utility economics is real planning and the advertisements, press releases, demonstrations, referenda, and so on are "down stream," part of the politics of adoption and implementation. Such a modernist interpretation might be appealing and comforting, but it would be precisely the wrong one to make. The heart of planning consists of inventing alternative courses of action and shaping expectations about what alternatives deserve serious consideration. Thus the interesting story from 1985 through 1989 concerns the turn away from the constrained path of public utility economics and into the complex world of Chicago politics, a world in which all participants sought to shape the course of the City's exploration in accord with their own diverse criteria. Planning in the "scientific" sense did not "disappear"; it became part of the rhetorically complex concept and activity that is planning.

So, most of the abnormal discourse recounted in this chapter expressed and reconstituted Chicago's mix of passion, reason, and power. What becomes of planners (in the sense of professionals who occupy jobs labeled "planner") in such an abnormal discourse? The first thing to note is that they seemed to fall out of (or disappear from) the Chicago story as much as modernist, scientific planners did. Many people and organizations influenced the flow of the City's exploration, but only one of them—the Director of Energy Management for the City's Department of Planning—was educated as a planner or occupied a job labeled "planner."

That one individual, about whom I have said very little to this point, deserves special attention. Recall that Mayor Washington asked the City's Planning Department to conduct a one- to two-year study of the City's electric power options. That task was assigned to a small group of energy planners who worked for the Director of Energy Management. Those planners did not try to prepare a "scientific" analysis of the City's options. They sponsored Beck's report. Nor did those planners try to identify some broad set of goals and objectives that all Chicagoans could agree to. They allied themselves with the energy efficiency and economic justice objectives favored by the community and consumer groups. Nor did they act as sterilized laboratory technicians who wanted to avoid contamination by the dirty world of politics. They identified themselves with Mayor Washington's political agenda and sought both to influence it and to bring it to fruition. The City's energy planner and his staff, in other words, drew on Beck's policy analytic research (which in turn was based on the normal discourse of public utility economics), promoted the values and interests of energy conservation and neighborhood development groups, and were part of Mayor Washington's pragmatic reform agenda.

The Department's energy planner did not act as a scientist, advocate, or politician. Nor did he act as a policy analyst, advocacy planner, or political entrepreneur or otherwise try to achieve some grand synthesis. Rather he sought to link Mayor Washington, the community coalition, and energy scientists by *mediating* (that is, understanding and translating) their diverse rhetorics. He did not mediate in a neutral fashion, however, as if he did not care what the City did with regard to the Edison franchise. He *actively* mediated the discourses: he personally identified with the Mayor's pragmatic but progressive reform agenda, agreed with the community coalition's desire to encourage energy conservation and local empowerment, and knew enough about the technical literature on electric power planning to stick within the realm of the possible. He accepted that each actor's claims were legitimate and worthy of consideration, but in order to ensure that each claim was fairly considered, he had to place Com Ed's interests on the periphery and to support the advocacy coalition that Edison sought to de-legitimize.

Was this Chicago planner acting ethically? In an enlightening article about normative ethics in planning, Elizabeth Howe (1990) creates a fourfold typology of normative theories concerning planning ethics (act-deontological, rule-deontological, act-teleological, and rule-teleological), and then observes that the presence of these four theories produces "ethical pluralism" in planning. "Each group thinks its own perspective is obvious," she writes, "and that of the others is basically flawed. Each can justify its position, but neither can convince the other" (128).

Using Howe's terminology, I would characterize the rhetoric of the scientist and the politician as being teleological; that is, concerned with the goodness (or utility) of an act's consequences. Scientists (public utility economists, for example) seek to predict or explain likely consequences, whereas politicians (such as Mayor Washington) seek to produce desired consequences. Similarly, I would characterize the rhetoric of the advocate as being deontological; that is, concerned not with the consequences of action but with the rightness of the acts themselves. Edison, for example, stressed the rights associated with ownership. Conversely, the community coalition stressed the democratic right to have a say in public decisions. So, drawing on Howe, one might conclude that when planners speak the rhetorics of science or politics, they are also speaking in terms of teleological or consequentialist ethics; when they speak the rhetoric of advocacy, they are also speaking in terms of deontological ethics.

What then of the Department's planners who tried to *actively me-*

diate among the diverse rhetorics of passion, reason, and power? What ethical theory were they articulating? My sense is that such a planner can be characterized, again in Howe's terms, as "a moral agent balancing a number of ethical principles" (147) in specific places (Chicago) at specific times (1985–1989) on specific issues (an exploration of electric power options) in a context of rhetorical and ethical pluralism. Planners who actively mediate reject the notion that planners can objectively determine what is right and wrong or objectively calculate the good and bad consequences of alternative courses of action. Conversely, such planners would also reject the notion that "right" and "good" are purely relative concepts. The notion of active mediation in a context of rhetorical and ethical pluralism rests not on objectivity and relativism but between them. Thus, such planners seek to create, sustain, and participate in a public, democratic discourse that enables them (and others) to argue persuasively and coherently about contestable views of what is good, right, and feasible. They must discursively redeem their acts, at least in principle, before relevant audiences.

I argue, therefore, that between 1985 and 1989, the City's exploration of alternatives to remaining on Commonwealth Edison's system moved from the normal discourse of public utility economics into the abnormal discourse of passion, reason, and power (Chicago-style), and that the City's energy planner (quite ethically and with a modest degree of success) sought to actively mediate among those diverse rhetorics. He helped the City find a path through the dense foliage of Chicago politics. Though reasonable, this conclusion feels incomplete. The sense of incompletion stems from the City's stated intent to *explore* alternatives and the planner's effort to mediate that exploration. To explore means to discover and map what is already out there. It implies that the City's alternatives to Edison already existed and simply needed to be identified and measured. The difficulty with this trope is that it denies the transformative and generative power of language, and it locked the Chicago energy planner and others into rhetoric that simply continued the community and culture of Chicago politics. In this constrained view, the path was already there. Their challenge simply was to find it.

There was, however, one notable exception to this metaphor of exploration, and it draws our attention to how rhetoric can constitute a new community and culture. Recall Ralph Cavanagh's *compact* (a short-term franchise that made future renewal conditional upon Edison's performance) and the Chicago Energy Commission's innovative sale/lease-back proposal. If acted on, these generative ideas might have turned the City's exploration into an inventive reconfiguration of Chicago's future possibilities. Clever and creative, these generative ideas might have alleviated the fear that precinct captains would operate Edison's nuclear plants by assigning managerial responsibilities to a private firm. They might have alleviated the threat of a tidal wave washing the City's economy out to sea by assigning ownership to a public corporation. And they might have ensured public accountability by establishing a contract that specified expected performance. Together, they might have blurred the distinction between public and private and created the possibility that the City could stimulate energy conservation and the growth of alternative sources of energy supply. A new path.

Though clever and inventive, these generative ideas lacked a complementary rhetoric. Instead of being presented as part of a persuasive story that the people of Chicago could understand and embrace, those ideas appeared as obscure, complicated, and probably untrustworthy parts of the stale old rhetoric of Chicago politics. Consequently, it appeared unlikely (at least in early 1989) that the sale/lease-back proposal would prove persuasive. Besides, it was lost in the tidal wave of worry about precinct captains, as can be seen by turning to a close analysis of the CACI's survey of Chicago businesses.

Interlude

A Community Group's Storefront Headquarters

The Chicago Electric Options Campaign's office is located twelve to fifteen blocks northwest of City Hall and the Loop's gleaming towers. I wish I knew more about the neighborhood, about its history and the novels that have been set here. Al Capone? Studs Lonigan? As an urban planner, I'm struck by how the neighborhood is in serious disrepair, decaying physically like much of the city, and by how different it looks and feels from the shiny new "high-tech corridor" that stretches into the western suburbs of Chicago. Walking toward the CEOC's building, I see many other buildings for rent. Most are older two- to four-story brownstones. There are vacant spaces and small parking lots, linked together by streams of trash and shattered parcels of pavement. The CEOC's building sits near a busy intersection. It is a one-story yellow stucco building with newer windows. Cars and buses roar by, spewing exhaust and kicking up dust. I walk in and discover that the ground floor is occupied by National People's Action. I see mostly white and Hispanic people, dressed in jeans and other casual attire. I walk down to the basement, where the office of the CEOC's one staff member is located. It all looks very temporary, but with ample room for a large number of people to gather. A place for organizing.

8

Most of the time I have to back up my numbers with reality.

— AN EDISON-SPONSORED SURVEY RESEARCHER

Survey Research as a Trope in Electric Power Planning Arguments

Earlier in this book, I argued that planning tropes are used in particular contexts (places, times, and issues) and thereby help give meaning and power to the particular narratives of which which they are a part. Earlier chapters showed that both Com Ed and its opponents had by the mid–1980s developed conflicting stories about the meaning of and need for Edison's last three nuclear power plants. The preceding chapter also revealed how those conflicting stories became part of the City of Chicago's effort to explore alternatives to Com Ed's electric power system. Those stories interweaved with one another in a complicated dialogue.

In this chapter, I want to focus on one seemingly small part of that dialogue. By the middle of 1989, the attitudes, expectations, and intentions of Chicago businesspeople had become a potential turning point in the flow of replies and counterreplies. To retain ownership of its system in Chicago and its franchise to serve customers in that city, Com Ed needed to persuade the businesspeople and citizens of Chicago that a City *takeover* of Edison's system would wreak economic havoc. To accomplish this objective, Edison needed a particular planning trope: a survey that claimed to measure objectively the attitudes, expectations, and intentions of Chicago businesspeople. Edison's opponents interpreted the survey as an effort to manipulate public perception, but they were unable to question the researcher about the survey in public until the middle of 1989. Then, just six months before the franchise was due to expire and as a Task Force appointed by the Mayor of Chicago was deliberating what the City should do, they finally had their chance to confront the researcher.

The Flow of Contending Stories Briefly Recalled

Let me briefly recall the flow of the contending stories as they interacted in the latter half of the eighties. Edison finished building its last three nuclear power plants eight years behind schedule and $11 billion over budget. Edison and its critics explained the delays and overruns by two very different stories. Defining themselves as expert managers of an extraordinarily complex enterprise, Edison officials attributed the delays and overruns to inflation and other factors *beyond its control,* especially excessive government regulation inspired by politically motivated interest groups. Accordingly, the company sought and obtained six substantial rate hikes during the first five years of the 1980s, including an 11 percent rate hike for its Byron 1 unit in late 1985. Then in early 1986 Edison announced that it would need another 32 percent rate increase to cover the cost of its last three nuclear units (Byron 2 and Braidwood 1 and 2)—unless, that is, some nontraditional approach could be found.

Consumer groups told a very different story. In their view, *managerial incompetence and profit seeking* had caused the delays and overruns. Edison's managers had built nuclear plants that were not needed, made northern Illinois one of the highest cost electric power regions in the country, and driven jobs and businesses away. Accordingly, consumer groups petitioned the ICC to cancel the Braidwood units, appealed the Byron 1 rate hike to the Illinois Supreme Court, and strenuously opposed Edison's efforts to obtain a rate increase for its last three plants.

The Mayor of Chicago, Harold Washington, joined the consumer groups in opposing Edison's rate increases. He feared that Edison's rate increases would harm the city's economy; he especially worried about the jobs and businesses that would be lost as a result of Edison's high rates. Motivated by this fear and his desire to pursue a progressive reform agenda, Mayor Washington took advantage of Edison's role of *servant* to the City and its residents pursuant to a franchise agreement that was due to expire in December 1990, unless jointly renewed by Edison and the City. In October 1985, therefore, the Mayor directed his planning staff to explore options to remaining wholly dependent on Com Ed.

In August 1986, a City-sponsored report estimated that the Chicago area would lose 85,000 to 112,000 jobs over the next twenty years if Edison received a 32 percent rate increase to pay for three nuclear power plants that were nearing completion (Ziemba 1986). A year later a second City-sponsored report estimated that Chicagoans could save $10 to $18 billion at a cost of $1 to $7 billion over a twenty-year period if the City bought and operated a portion of Com Ed's facilities (Beck 1987).

Edison attacked the Beck report vigorously. One particularly interesting attack came not from Edison directly but from one of its allies, the CACI. Roughly three weeks after release of the Beck report, and shortly before the Chicago City Council was to vote on whether to fund an additional $500,000 to study Beck's proposal, the CACI released the *results* of a survey of Chicago businesses that had been conducted by Michael McKeon and Associates. According to Sam Mitchell, president of the CACI, the survey showed that nearly 30 percent of the 454 firms surveyed would relocate outside of Chicago if the City became the sole provider of electricity, thereby costing the city 250,000 jobs. "We think," Mitchell said, "it would be prudent for Mayor Sawyer and the City Council to think long and hard about whether it is worth spending an additional half-million dollars to study a proposal which sends a negative signal to the business community. . . . The government must realize that the moment a half-million dollars is authorized, they set in motion planning by the business community that could ultimately cost Chicago a quarter of a million jobs" (Recktenwald 1987)

Consumer and community groups, like the Chicago Electric Options Campaign (CEOC), condemned the CACI's survey as "sheer propaganda." A year later they placed referenda on the ballot in fifteen wards. The vote showed extremely strong support for the view that the City should seek lower electric rates by exploring alternatives to reliance on Edison.

In early 1989, Mayor Eugene Sawyer, who had replaced the recently deceased Mayor Washington, appointed an Energy Task Force to help him decide what to do about renewing Edison's franchise.[51] On April 13, 1989, that Task Force listened to a lengthy presentation by Edison officials. There the Task Force heard a vice president of Com Ed say that the CACI survey showed that the people of Chicago wanted Edison to be their electricity supplier. Three weeks later, on May 2, consumer groups spoke to that same Task Force. Sam Mitchell of the CACI was the second speaker that day. Among other things, he reiterated the results of McKeon's 1987 survey of businesses. A spokesman for another community group immediately criticized the Task Force for allowing Mitchell to speak at that time. He claimed that Mitchell had merely reiterated Edison's position, and he urged the Task Force to "consider whether or not you don't have a Trojan Horse in your midst" (Mayor's Task Force on Energy 1989).

The CACI's survey was, therefore, part of a flow of utterances and replies pertaining to the City's exploration of options. By the middle of 1989, the economic effects of alternate courses of action, most notably a possible purchase of part of Edison's system, had become a focal point for the Mayor's Task Force. The CACI claimed that McKeon's survey truthfully showed that vast numbers of Chicago businesses would leave if the City "took over" a significant part of Edison's system. Consumer groups challenged that claim and sought an opportunity to question McKeon and his client. They finally had their chance when Sam Mitchell and Michael McKeon accepted an invitation to meet with the Task Force and other interested parties on August 16, 1989. Mitchell and McKeon would have an opportunity to persuade a key audience that the survey had been conducted in a valid manner, and consumer groups would have a chance to test their challenges to that claim.

The Energy Task Force's Response to a Survey of Chicago Businesses

As the meeting began, twenty or so people sat at the outer side of a hexagonal pattern of tables.[52] On the left sat several Task Force members; on the right, community and consumer group representatives. Another ten to twenty observers were scattered around the perimeter. Robert Wilcox—a conservatively dressed, graying, middle-aged businessman who cochaired the Mayor's Task Force—called the meeting to order, then asked the consumer representatives to comment on a written question from the Task Force. For the next hour or so, they engaged in a calm and reasoned conversation—one well-suited to the plain, unadorned, modernist room in which the meeting took place. The setting was rather typical for the conduct of modern planning.

Wilcox then interrupted the flow of the meeting to give Sam Mitchell of the CACI a chance to speak. Hunched forward, arms crossed, Mitchell read the following statement:

> Thank you, Mr. Chairman. We appreciate the courtesy. The Task Force in a directed question asked CACI, "Would you provide your survey instrument and the tabulation of results? What sources of information did you use to develop that instrument?" I believe that we have answered that question before, both to this body and to others, and that answer is "no." Our rationale is

that our research is proprietary and we do not release proprietary information. We release the results of our surveys and the questions concerning those results and the methodology used in conducting that research. That has been the policy of the Association since 1904; it remains the same today. While many in this room will not find that answer to their satisfaction, that is the reality of the matter.

Another reality is that our research appears to be the only document that has been brought to this body that has been impeached in this manner. To our knowledge no other group's work has been challenged. We believe that in . . . challenging the credibility of this research and the Association's credibility is placed in question. We can only draw the conclusion that some members of this Task Force did not like the results of our work and seek to discredit it in order to feature their own aims.

As I said a moment ago, we have always been willing to answer questions concerning the results of our research and the methodology used in conducting that research. Toward that end we have asked Mr. Michael McKeon, of Michael McKeon Research Inc., who conducted this research, to appear here to discuss the methodology and the results. Mr. McKeon is well-known and highly regarded in the field of market research and is widely used by business in the region. I feel obliged to point out that Mike came back from vacation on the East Coast to meet with the task force and answer these questions and, if it pleases the Chair, I would like Mike to join us at the table.

The meeting did not, therefore, begin well. Mitchell indicated both that he felt the credibility of his organization (and the research it had sponsored) had been unfairly challenged and that he had no intention of giving the Task Force what it was asking. But he did give them Michael McKeon.

McKeon began by indicating that he was unhappy about having to interrupt his vacation in order to speak to the Task Force. Then, interrupting a questioner, he said:

Well basically what the process was we were requested to do was do a survey of businesses in the Chicago area to see what their feelings was on a study that came out two years ago, the initial one, which called in some cases for I'm not nearly an expert on the stuff which I'm not expected . . . I'm expected to test about the public opinions. Basically what we did was, is we had Dunn and Bradstreet generate in three divisions for us random names of companies located in the city of Chicago, based on fifty and under, fifty employees and under, fifty-one to two hundred fifty employees, and over two hundred fifty employees. We designed the survey to just basically test their mood on what they felt was the, the results have been released here, on what they felt the prospects of the city running the, having something to do with running, the

> taking over the electric service, providing electric service and things like that. The research was done under controlled means, statistical probabilities and everything else like that. And, we just ran it. We designed the questionnaire so that we could, so that we got a free flow of information.

At that point McKeon shifted topics, angrily portraying himself as an expert whose work had been criticized by politically motivated know-nothings:

> I might point out that, just for the comments I've heard here, it's interesting you know, it's the common problem I have with clients. I might point out that our firm was the first one in the country to track the LaRouche problem long before it ever it happened. . . . We were the first ones five years ago to release our findings nationwide on crime and drugs as a major problem in the country which everyone else ignored. . . . [I]n the governor's race of Stevenson versus Thompson we were the only one to hit it accurately. The last time around in the mayor's race we came out with the numbers far before anyone else in the public and was proved correct.
>
> Basic, what my business is, which is a lot different from most of the market research firms, is most of the time I have to back up my numbers with reality. The elections come, the elections go, the referendums happen and go down; [pounding his hand on the table] my numbers are in the papers ahead of time so that they're checked and they're right. And quite frankly from what I've heard questioned here, [voice rising] the only questions ever heard, the only ones that have questioned, the last three, four people have questioned my numbers. . . . Basically what happens is is when people don't pay attention they don't they don't like the results, whether it's really reality or not. And my numbers have always been that way. Yeah.

So the issue according to McKeon is that he has had predictive success but his critics (many of whom are in his immediate audience) are too ignorant to know it. Furthermore, the quality of his work can be assessed by comparing his preelection surveys with actual election results. At that point one of the community representatives—Scott Bernstein, Executive Director of the Center for Neighborhood Technology—asked a question. A thin young man with short, wiry black hair and beard, Bernstein wondered whether there might be a significant difference between a survey of business intentions and a survey of voter intentions. "[I]f that's the case," he said, "you were widely quoted in the same newspaper, or your survey was by CACI, as having something to say about the likelihood of businesses' intent to leave should the city act to play some sort of municipal utility role. Would you comment for us on the nature of that question [here McKeon tried to interrupt] and what and how and how you approached it and whether you feel as strongly about that as you feel about the likelihood of the outcome of the Attorney General's race?"

By this time McKeon had become extremely hostile. His anger, coupled with the CACI's decision not to present the actual survey instrument, led to great confusion in the meeting. The confusion concerned the precise wording of the questionnaire and whether that wording had skewed the results in Com Ed's favor. Dodging Bernstein's question about comparability of surveys, McKeon focused on whether businesses would be likely to leave the city:

> Oh I think there would be significant shifts. I think there were two parts to that question. One . . . part of that proposed survey called for 50 percent cogeneration, if you recall[53] . . . [pointing at Mitchell] Am I right or wrong? [Then after hearing an affirmative answer from Bernstein:] Fifty percent cogeneration is . . . not what . . . we threw into the action, but what the report that your study and analysis [the Beck report] said was there. Fifty percent cogeneration for the businesses. . . . Are you familiar with that? It's in the study. It's in the Beck study. [Angrily jabbing his finger toward Bernstein] True or false? Are you familiar with that?

At this point McKeon's barrage of hostile comments was interrupted by one of the Task Force members, Michael Bell of Certified Public Accountants for the Public Interest. Bell sought to bring the conversation back to the initial question. "I haven't made any comments pro or con," Bell said, "but I feel like you're getting in here and boxing with somebody. . . . Could we just talk about . . . what the issues are here rather, because it just feels to me . . . that this is getting pretty loaded and kind of off the tangent. . . . What I am interested right now is in process and in some of the questions the way they were asked. . . . We're not getting . . . the survey instrument so we're, we want to be able to rely on what you are saying."

After another brief flurry of heated exchanges, during which McKeon referred to the Task Force as a "hostile committee," the conversation returned to the 50 percent cogeneration issue. Martin Heckman, an informally dressed older man who represented the Labor Coalition on Public Utilities, asked thrice whether the survey instrument could be made available. Finally, Sam Mitchell of the CACI intervened, saying impatiently, "I'd like to answer that if I may. Mr. Heckman, at the last meeting I saw you in fact carrying a copy of an editorial from *Commerce* magazine which outlined in totality . . . the results of the survey. You have that material in your possession." McKeon then chimed in: "The questions that are on that survey are the questions we asked. Period. There was no other loading, there was no other bias or anything else like that. And so, what you see is what you got. The survey instrument is exactly what's on the poll. Yes sir [gesturing to the next questioner]."[54]

The Task Force then inquired about the actual phrasing of the questions. David Kraft of Nuclear Energy Information Services, a public interest organization that sought to inform the public about the risks associated with nuclear power, opened the inquiry: "What sort of reliability checks were done on the terms used and the questions, for example, the words *city, became,* and *sole provider,* to make sure that that was consistently interpreted from participant to participant in the survey?" McKeon had some difficulty answering Kraft's question: "There were . . . other questions of supply," he said. After McKeon then rambled a bit about how the survey had been conducted shortly after Mayor Washington released the Beck report, Kraft responded that "essentially what you're saying is, the only estimate of reliability was the context of what was going on at the time?" Though calmer now, McKeon still did not seem to understand. "All right," said Kraft,

> I'll explain more clearly if I can. If that question were given to me now, as opposed to then . . . You see it's a different context first of all. And secondly . . . does the term *city* connote to those participants the aldermen as have been portrayed in the press, did it connote the Department of Planning, did it mean city hall? . . . I mean . . . there are different ways of interpreting it is what I'm what I'm driving at. And the same with the word *sole provider,* did that mean that they were the ones who are actually going to produce the electrons or did it mean they would be the fiscal backers or the, see what I'm driving at, you see what I'm driving at?

McKeon seemed to understand. "First of all," he said,

> what I think you have to look at . . . is . . . you're . . . an energy task force that's looking for this, looking for that. What we deal with is public perception. The public has a very limited perception of what's going on. You can talk about energy supplies, Mr. Bernstein talked about one of his top client Sears moved out of Chicago now, moved to, moved right outside the city where the rates are still the same and everything else like that. . . . When they think of *city,* what do they think of? Any kind of municipalization. Now we could have been more negative, we could have said that the Chicago city council is considering taking over the collection company. Right through the roof. My expertise, what I do best, I feel I do it better than anybody else, is is we try and

make the questions as neutral as possible so what we can get out of it is the people's interpretation of what's going on, not what the clients feel is there or anyone else for that matter.

Kraft chose not to challenge the validity of McKeon's survey results. He did, however, repeat his worries about reliability: "What I'm driving at," he said, "is I needed a better clarification of the reliability of the particular instruments because you said, you say right now, what the name of the game is: it's public perception. . . . Depending on how these words were perceived, people will frame their answers." Undaunted, McKeon simply claimed that "they were highly neutral words."

The discussion then turned to another question that was, in the eyes of some Task Force members, "loaded." One of the consumer group representatives asked McKeon whether the survey asked a question that indicated the City would *require* businesses to reduce their electricity usage by 50 percent, either through reduction in demand or by replacing utility power with cogenerated power. McKeon's answer was difficult to follow:

We did the clear first, with the city of Chicago becomes a becomes a sole supplier of your electrical supply, how will this impact your business in your planning, were the exact terms. OK? After that, we said well what if they had to come up with cogeneration and things like that and conservation and things. And then the numbers went right off the chart. [Then, when asked again whether the question indicated that a 50 percent reduction would be required,] . . . What if you relied on 50 percent cogeneration? Yeah. . . . That type of thing.

At last the conversation had turned to something concrete. After scanning through some papers, Scott Bernstein zeroed in on the 50 percent figure. "Just a point of clarification," he began:

The scenarios that were provided by the R. W. Beck Company for cogenerated power [thumbing through the Beck report] did look at the potential of 2,100 megawatts of cogeneration but the figures actually used in the various scenarios are [unintelligible] no, no way. We're talking about a base load of 4,500 megawatts at a range from 35 to 800 from cogeneration, a peak of 1,322 megawatts for conservation and cogeneration alone in the year 2005, when demand would have increased by almost 1,000 megawatts also equaling roughly what we're projecting here, something like a 20 percent contribution from, in effect, improvements in technology. So if you ask the question you know, would you, what would you do if you had to have 50 percent cogeneration, could you . . . [at this point both Bernstein and McKeon speak at once, and neither can be understood]. . . . Here's the study. You can read it yourself.

McKeon seemed surprised and a bit shaken. "Is there 50 percent?" he asked. "Could that be 50 percent? No? Never could be interpreted so? With the cogeneration it never could be interpreted as 50 percent?" Sam Mitchell stepped in at this point and tried to shift responsibility for the erroneous question to the City. Yet then McKeon shifted ground:

> But this is specious because of the fact that the question before that without even without the cogeneration still had a ton of people moving out, and what were the numbers on that? It was still there, it was still a negative impact. . . . You can sit and shake your head. . . . If we had asked 20 percent, [gesturing angrily again] . . . then they'd still have gone. Are you kidding me? You ask any company, do you own businesses around here, any of you own businesses? The city comes up and says, oh, by the way, you might have to do 20 percent cogeneration, what do you think the results are going to be? Do you think the results are going to be any different at 20 percent than 50 percent? [Jabbing his finger across the table at Bernstein:] Ask him. You are the one who thought of it? Do you think it's going to be any different at 20 percent than 50 percent?

Unimpressed and unpersuaded, Bernstein tossed his papers onto the table. Shrugging, he simply said "I don't think that's the question. You've answered the question that's on the table, how you conducted the study and what your assumptions were. Thanks a lot. It's self-evident."

At that point the meeting appeared to be over. But then, after two or three minutes of side conversation, one other key issue returned: the credibility of the CACI and McKeon's research. Sam Mitchell stressed that he had asked McKeon to take part in the meeting because the survey research had been "impeached." That offended Mitchell: "We know of no other document brought before this body that has been impeached. We consider that not only a question of the credibility of the research and the professionals we hire, we consider it a question of credibility of the organization. We . . . " At this point Michael Bell began objecting while Mitchell was still speaking. "The Beck report was impeached, Mr. Mitchell, too," Bell calmly insisted. "This is an open society. The Beck report was impeached on a number of different points. There is no reason why there should be any higher ground for a survey and the questioning of what the survey is . . . " "Our point," Mitchell replied, "is there should be no lower ground." Bell found this response to be deeply irritating:

> Mr. Mitchell, I'm not going to let you get away with saying that no other report was impeached. That's incorrect. You are doing a disservice to a number

of the members of this task force by . . . assuming that . . . a survey . . . that you conducted and so on is the only thing that is being impeached. That is not correct. . . . The body of the task force and a substantial number, I don't know if it's the majority, have a concern to try to be on an equal ground here, on a level ground. There may be people who have their biases, we come into life with biases, that's the way it is. But to say that a survey is the only instrument is misstating it. The Beck report was impeached, you've done it yourself. Let's get beyond that and see where we are. . . . I would very much appreciate having people in the arena to work some of this out rather than getting back into this this business of them and us, with due respect. . . . It bothers me completely to have this gentleman come in [gesturing to McKeon], I respect your process I have no reason to doubt that process, to come in and box with everybody like we've all got the same maps. Let's stop it.

Sam Mitchell tried then to indicate that the Task Force itself was biased and that its reaction to the survey reflected that bias. "I, in my exposure to this body," he said, "have heard this research impeached. I have heard no comments as to the Beck study, in front of this body when I have been here, and I have seen nothing of any importance on the Beck study being impeached by others. I bring to you, to this body, then, my biases to what I understand which is a severe question of the credibility of the product that we produced and the credibility of the organization. That's why I've gone to the process of bringing Mike in." Disturbed by use of the word "impeached," McKeon made one last comment, addressing Bell: " You said that this isn't the only instrument that's been impeached. I don't think my stuff has been impeached. You said the Beck study has been impeached; there is no reason to say why this study is. Do you think this study has been impeached? . . . The other thing is, too, is that is if the committee is going to operate on fair ground then the members of the committee shouldn't be taking shots in the papers about stuff that you don't even know."

McKeon then began to leave the meeting. But before he left, Martin Heckman of the Labor Coalition made one last comment. "Bob, I'd like to make one point. The point is the Beck report itself is impeached. The day that the Beck report was released, on that very day within a matter of an hour, Commonwealth Edison threw somebody in to raise the question about the city wants precinct captains to run Commonwealth. So they tried to impeach it within hours."

Planning as Persuasive and Constitutive Discourse

In chapter 2, I claimed that planning is a rhetorical activity, and that planners should think of survey research and other tools as rhetorical tropes that reply to prior utterances (and give meaning and power to the larger narratives of which they are a part), seek to persuade specific audiences, create open meanings subject to diverse interpretations, and help to constitute planning characters and communities.

The case just recounted strongly supports such a rhetorical conception of planning. *It shows first that the CACI's survey acted as a trope in Com Ed's narrative about the desirability of the City "taking over" Edison's electric power system.* The survey replied to two prior reports prepared for the City, reports that indicated Chicagoans would do better if the City purchased all or part of the Com Ed system. How Chicago businesspeople felt about (and would react to) such a purchase was unknown. The survey, presented in the scientific terms of "results," supported a claim that the businesspeople would be deeply displeased. But the claim that the survey simply measured attitudes and intentions belied its tropal nature. Given the businesspersons' "limited perception," acknowledged by the survey researcher himself, the survey had to construct their understanding of the situation. Thus it had to prefigure their sense of how they would respond.

Here is one particularly revealing example. McKeon's survey read: "Part of the city's plan for the takeover of the electrical service would require a business to reduce electrical consumption by 50 percent or generate half the electrical power it uses. Under these conditions, what effect would this plan have on your company's future plans?" (Anonymous 1988). Looking at the wording of this question carefully, we can see how three particular aspects of this question functioned as figures of language or argument and prefigured how businesspeople would respond.

First is "the city's plan." In context, this phrase seems to refer to the Beck report, released a few weeks prior to the survey. Neither the City nor the Beck report, however, had referred to the report as the city's "plan." The City portrayed the report as a quantitative assessment of the likely economic consequences of alternative courses of action, not as a plan about what to do and how to do it. Indeed, the CACI's release of McKeon's survey occurred just a few days before the City Council was to decide whether to fund the additional studies required to help the City decide what to do. To refer to the Beck report as "the city's plan" probably was to convey the misleading impression that the City already knew its intentions.

Second is the claim that the city's plan would "require" businesses to conserve or generate 50 percent of the power they use. As Bernstein's questions tried to demonstrate, nothing in the Beck report or in the City's actions up to December 1987 indicated that businesses would be "required" to do anything. The City was clearly inclined to encourage electricity conservation and cogeneration and to do so by offering price and other institutional incentives, much as Commonwealth Edison was then offering numerous incentives for businesses *not* to conserve or cogenerate. But at no point had the City indicated an intention to "require" businesses to conserve.

And third is the survey's use of the term "takeover." This word implied that the City was on the verge of seizing something that belonged to someone else, that it intended to expropriate property. This trope radically transformed the context and meaning of the City's actions. Though the survey never said so, the City's exploration of options was taking place in the context of an expiring franchise. In 1948 the citizens of Chicago had granted Edison the exclusive right to provide electric power and services to the city in return for payment of an annual franchise fee and other services provided to the City. That franchise agreement was due to expire in December 1990, unless jointly renewed by Com Ed and the City. Thus the City had every legal and moral right to explore options to remaining on Com Ed's system and to continuing the exclusive franchise. To portray this simply as a "takeover" was to twist and impoverish the meaning of the City's actions in an objectionable way.

To be sure, the CACI and Mr. McKeon might want to defend the accuracy of such terms. But the issue here is more their figurality than their accuracy. Planning and its instruments must use these kinds of words. Yet words like "the city's plan," "require," "takeover," "proprietary" can coalesce to make any survey into a powerful trope of argument—not just because such words might mislead readers, but because they shape our understandings of the situations at issue. By using a particular language, the survey made "the city" into a synecdoche, standing for the board of aldermen and the political machine in Chicago. By contrast, community groups used "the city" as a metaphor, meaning *the city is us.* One can imagine a case for either construction, so my main point is not that one side is purely right and the other damnably wrong. Instead the key lesson is that no planning instruments can avoid persuasive, rhetorical construction of characters and communities. Scientistic talk of survey "results" might obscure that crucial feature of planning, but cannot eliminate it. The same tropality appears in the survey's division of "the business community" from "the city." Surely this joined with "takeover" and other terms to help survey respondents think in terms of corrupt politicians expropriating private property and forcing private businesses to do what the politicians wanted. Would many businesses leave Chicago under such circumstances? What do you think? But remember that there can be no neutral words about contested issues: other survey phrasings would lead in other directions. The lesson is not to avoid or debunk surveys, but to understand how they must work tropally. Then we can do them well and read them wisely.

The case also shows that the survey researcher and his client had to persuade a specific audience (the Mayor's Task Force) that the survey was conducted competently and that its "results" were accurate and reliable. When the Mayor's Task Force began meeting in early 1989, it seemed as if that Task Force would play a crucial role in determining what the City would do with regard to its franchise with Com Ed. The economic impacts of alternate courses of action were a crucial factor for it to consider. The utility had to persuade this specific audience in this specific context that the economic effects of a City "takeover" would be disastrous for the City. The CACI's survey and Mitchell and McKeon's subsequent meeting with the Task Force were efforts to persuade the Task Force how great the disaster would be. Whether they succeeded in those efforts, however, depended in large part on the extent to which the survey acknowledged and fairly incorporated opposing views held by (or familiar to) Task Force members.

The case also shows, therefore, the importance of speaking with awareness of differing or opposing views, and how particular audiences can—due to the inherent plasticity of language—read meanings into surveys that differ from the one intended by the survey's client or researchers. The CACI's survey displayed no overt awareness of opposing views, and suppression of that awareness became quite evident during the Task Force's meeting. Intimately familiar with the flow of utterances, replies, and counterreplies of which the survey was a part, several members of that Task Force read the CACI's survey in ways different from the CACI. They challenged the CACI's motivation in producing the survey, and they convinced themselves that the survey was simply an effort to deceive and manipulate the business community. "You've answered the question that's on the table," said Scott Bernstein—"Thanks a lot. It's self-evident." Mr. McKeon's inarticulate but indignant response might make us wonder about whether the situation was that simple. And it is interesting that the critics on the Task Force did not challenge the notion that surveys could produce objective "results." Their implication was instead that the survey asked the wrong ("loaded") questions. They believed, perhaps with good reason, that other questions—possibly ones based on the views of community groups—would have yielded starkly different, likely better responses.

One might think, therefore, that good survey researchers could have obtained a "true" measure of the business community's response by testing the sensitivity of responses to language. That is the technical fix taught to modern planners, and though not completely incorrect, it typically forgets that any planning survey must figure as a trope of argument. At another pole of the postmodern arguments about planning, we might postulate a survey conducted in the context of a Habermasian "ideal speech community." Then the CACI and the community groups would gather happily together with a planning researcher to choose questions that would test the sensitivity of business responses to diverse phrasings (Habermas 1987). One can imagine that such a revised survey could have gone a long way toward reducing conflict between Com Ed and consumer groups, and one might hope that creating such a planning forum based on principles of honesty and trust would have led to a better understanding of what Chicago residents think of a potential takeover. But it is important not to forget that task forces and surveys and the like exist precisely because they sometimes seem the closest we can get to such undistorted, democratic dialogue. Both the technical and the communitarian utopias are pleasant dreams, but dreams and figures nonetheless. Planning is, as its practitioners well know, a deeply politicized practice, and no technical fix or ideal speech community is going to overcome that.

Lastly, the case also draws attention to the ways in which planning rhetoric (in this case Michael McKeon's) constitutes character and community. The story reveals important differences between how to *do* survey research (do planning) and how to *be* a survey researcher (be a planner).[55] Most importantly, the tale encourages us to ask whether the meeting would have gone differently if McKeon had adopted a different rhetorical strategy. My sense is that McKeon's rhetoric created a character for himself that was radically inconsistent with the character embedded in Com Ed's basic story. Furthermore, his rhetoric reaffirmed the expectations of Edison's opponents and thereby reconstituted the community and culture of conflict between them and Com Ed. His presentation to the Energy Task Force, and its members' questioning of him, were part of the frozen embrace of electric power politics, Chicago-style.

Recall that Edison's story characterized its officials as expert managers who were being harassed and impeded by politically inspired know-nothings. When coupled with the CACI's description of McKeon as a highly regarded market researcher, and with McKeon's self-portrait as a scientist who uses "neutral" questions to generate "results," this would lead one to expect that he would speak as a scientist addressing an audience of politicians and lay advocates. As a scientist, however, McKeon might have been compelled by the ethics of his science to reveal much more about his survey.[56] Then his scientific ethos might conflict with the political nature of his commitment to the client, be it the CACI or Com Ed. McKeon, like planners in general, would have faced the tragic choice of liberal politics: to be right and do good (in his own view) or to get things done (in his client's view).

So Michael McKeon's rhetoric constituted a character for himself that was radically inconsistent with Edison's (and his own) story. He spoke and acted not as a scientist but rather like a representative example of a combative Chicago political hack. His dress (unkempt), his gestures (jabbing his finger toward questioners), his vocal intonation (loud, angry), his frequent lapses into grammatical incoherence, all contradicted his claim to scientific character. By acting as a political hack, McKeon led many in the Task Force to conclude that he was deceitful and untrustworthy. His rhetoric made Edison's opponents' story seem more persuasive: the utility was headed by arrogant incompetents who sought to manipulate the public and its elected representatives.

We can wonder why McKeon posed (metaphorically at least) as the "political hack" and why he came into the meeting exuding such intense hostility. Lacking the chance to ask McKeon directly, I can only pose some possibilities. Perhaps that is just the way he is, at least under pressure. Most of us have encountered—or have actually been—such characters at one time or another, so we cannot dismiss that possibility. A second alternative is that McKeon truly regarded the Task Force members as technically uninformed and politically motivated meddlers. One might infer this from McKeon's words and demeanor. But if he thought that when he entered the meeting, how did he retain the opinion by the time he left? Bernstein and Kraft asked him too many technically sound questions to be ignorant, and Bell was much too evenhanded to sound politically craven. By acting as a political hack, McKeon seemed to implore or expect his audience to respond in a similar way: to be hostile, aggressive, and evasive—and thus to reproduce a (Chicago) culture (of politics) based on deceit and manipulation. Contrary to McKeon's invitation, in any event, the Task Force members chose to be different characters and not to "be McKeon." Bell, for example, appears three times in the conversation: first, to ask a simple, straightforward question; then, to ask McKeon to "stop boxing"; and last, to protest Mitchell's claim that McKeon's survey was the only work to be "impeached." He did not allow McKeon to go unanswered, but he chose not to respond solely in McKeon's terms. A third, strategic possibility is that McKeon and Mitchell jointly agreed that the best defense was a good offense, and hence attacked their opponents on grounds of ignorance and politics rather than directly answering questions and objections from the Task Force. My suspicion is that this is a better characterization. Looming behind McKeon's research firm, and behind Mitchell's association, was the pervasive shadow of Commonwealth Edison. As a power behind the scenes, it seemed to have nothing to gain and much to lose by revealing the extent to which survey research could shape the attitudes and intentions of Chicago businesses and residents. If McKeon had responded openly to Kraft's plea to design a survey that tested the respondents' sensitivity to words like "city" and "sole provider," he might have placed the Edison monopoly at risk. But a final possibility, one that also rings with truth, is that Mitchell simply threw McKeon to the lions. After all, the key question became whether the Task Force would be allowed to see the survey instrument, and Mitchell left McKeon in no position to share it.

How, then, might McKeon have responded differently to his audience, and what effect might a different rhetoric have had on the meeting, his audience, and the conflict between Edison's and its opponents' stories? One possibility to consider is that, rather than appearing to be a cartoon of how to fail at persuasion using the mask of science, he might have portrayed himself as a polished deflector of uninformed questions from outside his realms of expertise, as a politically savvy policy analyst perhaps. Presuming rhetoric to be "mere words," he might have spoken in a more polished and dignified way—while remaining just as uncommunicative. My sense is that a more polished researcher could have obscured the central issue—the tropal (and perhaps "loaded") nature of his questions—by saying something like "The majority of our respondents would not fully grasp a technical explanation of the complexity of the Beck report," or "We've learned that trying to fully capture subtle distinctions among options in survey responses diminishes the validity of the response." The polished researcher could then have argued that his survey might have contained some unavoidable, but minor, distortion. By obscuring the central issue, such a polished researcher would have appeared to be less of a "hack," but still would not have persuaded many in his audience. To persuade them, he would have to be willing to discuss details about how the survey questions were selected and how the survey was administered, and he would have to be open to the possibility of modifying his questions to account for his opponents' views. This he was unwilling or unable to do.

A second alternative is one probably unavailable to McKeon but one that planners should strive to make possible: to promote and facilitate a public, democratic, persuasive discourse in particular contexts about what path should be followed into the future. Planners attentive to rhetoric would "listen" (in advance) to their audience's story, seeking to learn how that audience thinks and feels and what kinds of questions its members are likely to ask. Having discovered the grounds for the audience's claims, they would pay closer attention to the alternative meanings that their tropes would have for that audience. They would try to persuade their audience, fully aware that their planning tools configure planning arguments. And such planners would speak in full awareness that their rhetoric has the potential to create new communities and a new culture of interaction between their audiences and themselves. But they would not be naive; they would assume that their audience is also speaking persuasively. Hence they would know that political effectiveness often means withholding some information and feigning more legitimacy than they actually have (Kaufman 1990).

Let us imagine such a planner in Michael McKeon's position. Knowing Edison's story and listening to the consumer groups' story, a planner attentive to rhetoric might have concluded that the consumer groups would continue to obstruct Edison's effort to obtain rate increases and hinder its ability to negotiate a new franchise agreement with the City. Aware that terms such as *the city's plan, require,* and *takeover* have different meanings for the consumer groups than they do for Edison and hence configure planning arguments, she or he might conclude it was in Edison's self-interest to probe the figurality of those terms with its opponents. In sum, such a planner would encourage Edison to invent ways to constitute a new community between itself and its opponents and thereby invalidate or change the consumer groups' characterization of the company. Freed by this reorientation, such a planner might provide the Task Force and the consumer groups with the information needed for them to discuss the survey's methodology and the meanings of its "results."

We can imagine such rhetoric and can argue that planners should strive to behave this way, but in the end we find it difficult to believe that a planner in McKeon's shoes would be so free. Thus, we conclude that McKeon's actual rhetorical strategy was vitally important because it reinforced belief in the consumer groups' characterization of Edison and thereby helped to reconstitute a community and culture of deceit, manipulation, and costly conflict. The difficult challenge now facing planners, in Chicago and elsewhere, is to find better ways of listening and speaking even when such structures of power strive to constrain us—and to conceive how we can begin in fact to talk in those ways.

In the end, this story about the uses of survey research in Chicago teaches us that we planners are engaged in a thoroughly rhetorical practice. Accordingly, we should surrender any further pretense to neutrality, objectivity, and universal truth—such as the "true" measure of business response to a City "takeover." Surrendering the pretense to objectivity does not mean, however, that we should flee to the extreme of defining planning as just another form of politics gone amok. We should, instead, embrace persuasive discourse and political conflict—to realize that survey "results" are, like all alleged planning "facts," inherently tropal and contestable. They must be scientific *and* rhetorical, professional *and* political, because they—like all other planning tools—configure our planning arguments.

Interlude

An Alderman's Office in City Hall

Hurrying to the Loop again. Don't want to be late for my meeting with the alderman. I find myself musing about how easily I and others equate Chicago aldermen with political machines and corruption. Perhaps it was reasonable to equate the two during the Progressive Era. Perhaps in the long years of "Hizzoner," the first Mayor Daley, too. But now? I wonder. I walk down a dreary hallway, pass by the press room, and come to the alderman's committee office. It is much smaller, more crowded, and far less private or distinguished looking than I expected. Two young, well-educated, and energetic staff members share the space with the alderman. Their desks sit in opposite corners, with the alderman's between them in the corner away from the door. They seem to be starting the office from scratch, and one of the staff spends a "Kafkaesque" hour searching for stamps. The phone rings frequently, often with a call from the alderman's forty-third ward office, but also from the news media about the capture of an alleged rapist, about a beer garden, about fly ash dumping, about an energy conservation hearing, about asbestos removal. Bill calls about a ticket. People drop by constantly: Betty to talk about zoning for Children's Hospital, former campaign workers to chat, a friend from the press office to use the computer, John Hooker (Com Ed's lobbyist) to talk about a transmission line, Maureen Dolan from the CEOC to talk about the franchise and an energy conservation hearing, a fellow alderman who refers to CEOC people as kooks. People and events swirling around them, the alderman and his staff are "trying to get up to speed on the franchise."

9

The city's already wimped out.
All they want to do is bore
everybody to death and
cut a deal in secret.
— MAUREEN DOLAN OF THE CEOC

I'm no wimp.
— MAYOR RICHARD DALEY

Precinct Captains at the Nuclear Switch: The Mayor's Hand Turns up Empty

The preceding chapter probed the tropal nature of survey research in Chicago's electric power planning arguments. It concluded that the survey researcher's rhetoric during his meeting with the Mayor's Energy Task Force reinforced the consumer groups' characterization of Edison as being poorly managed and politically craven and that the researcher's rhetoric helped to reconstitute a community and culture of deceit, manipulation, and costly conflict.

This chapter returns to the City's exploration and to the abnormal discourse within which it was being conducted. From 1985 through 1989 (as expiration of Edison's franchise with the City approached), the City, the utility, community organizations, and business groups engaged in a contorted argument over that exploration. Their argument was, in large part, driven by economic worries: the City's about economic decline and competitiveness, the utility's about being able to pay for its last three nuclear power plants, the community groups' about the fairness of Edison's rate increases, and the business groups' about rates becoming a political toy. Though driven by such worries, the public argument was not just about the fairest or least costly way of producing and distributing power, given the existing institutional environment. It was also a debate about the institutional structure itself.

The institutional structure defines how individual citizens and rate payers relate to certain organizations (Commonwealth Edison, the Illinois Commerce Commission, citizen groups) and how these organizations relate to one another. As shown in chapter 3, a set of "working rules" currently defines the rights, duties, and liabilities of these individuals and organizations with respect to electric power rate making and certain other power decisions. Here I address three questions about those rules: Why were some citizens of Chicago no longer content to work within the given rules of the game and the given institutional arrangement, and why did they seek instead to change that institutional arrangement? To what extent were they successful, and what factors appeared to facilitate or obstruct institutional change? And what is (and can be) the role of planners in understanding or contributing to similar debates about institutional change? In the end, those change-oriented citizens succeeded in persuading the City to terminate its existing franchise with Edison and to notify the utility that the City might exercise its right to purchase part of Edison's system. However, they were not successful in persuading the City to create a new institutional structure or to negotiate a new franchise that was more favorable to the City and its residents. This lack of success, I argue, can be attributed in large part to the inability of the City's planners or the community groups to invent a story about the future that Chicagoans could find invigorating and persuasive.

A New Mayor Holds His Cards Close to His Vest

In March 1989, Richard Daley decisively defeated acting Mayor Sawyer and Alderman Timothy Evans in the democratic primary for mayor. A few weeks later he won the general election by an even more decisive margin. He had said little about the franchise issue in his campaign (other than his "Four Cs" pledge), so it was not clear what he would do. Like a poker player who holds his cards close to the vest, though, this new mayor gradually demonstrated that keeping others guessing (that is, not saying much and controlling information) would be a key part of his approach.

The first several weeks were, as with any new administration, a time of transition. Though gone, the preceding administration still influenced actions through—for example—the policies it had adopted, the staff it had hired, and the commissions it had appointed. Though in power, the Daley administration had to work with those staff and commissioners and to deal with problems and opportunities left to it by Mayor Sawyer. As far as the City's exploration of alternatives to Edison was concerned, Mayor Sawyer's Task Force on Energy (TFE) bore the brunt of this transition. It had already begun its work just a few weeks before Daley's election, and—not knowing what the new mayor would do—proceeded in accord with its prior schedule.

The Task Force held the first of its public meetings in mid-April, shortly before Daley assumed office. Six Com Ed officials, led by Vice

FIG. 9.1 **Richard M. Daley, Center, Campaigning for Mayor**

President Donald Petkus, presented a very professional and polished argument that the utility alone had the expertise, manpower, and electrical capacity to serve the City (MTFE 1989b, April 13). But few on the Task Force seemed persuaded, and Edison's answers to their questions probably did not help. Here are a few key examples. When discussing the cost of Edison's power, Vice President John Bukovski said, "There seems to be a perception that our prices are surging dramatically. This simply is not true." Task Force members were skeptical, though, because they knew that Edison's prices would have surged dramatically had it not been for the opposition of the City, CUB, and other consumer groups. Likewise when Josh Hoyt of CUB and Otto McMath of the South Austin Community Council asked why Com Ed seemed so afraid of greater competition, Vice President Petkus replied that "we certainly . . . do believe

Here Edison officials legitimately claim that their electric power system is quite complex and that they have the expertise needed to operate it. One might wonder, however, about the contrast between this claim and the company's overtly political efforts to persuade the public to avoid having bureaucrats and activists "take over" "your electric power system."[57]

in competition. There are some restrictions on how you can compete and what type of competition that I'm sure you and I would disagree on, but we are certainly in a competitive world and we intend to survive in a competitive world." Similarly, when asked why Edison provided so few incentives to promote energy conservation and cogeneration, Petkus claimed that Edison had "been in the conservation business since the late 1960s." Several other Task Force members asked why Edison was engaging in scare tactics (like claiming that "precinct captains would be running the power company" and that "extremists" were trying to take over the system). Petkus said that he did not know of one person who had said that it would be a precinct system and that "our position is that the City and the Task Force should explore all the alternatives." Patrick Giordano, sent by Mayor-elect Daley to monitor the meeting, seemed less skeptical about Edison's presentation than the Task Force. Claiming that Edison had presented "some useful information," he reminded reporters that "the only option the mayor-elect has eliminated is the option of city employees running the utility. . . . All other options will be explored." He also said that Daley planned to preserve the Task Force but might reevaluate its schedule (Kaplan 1989).

Two weeks later consumer and community groups presented their case to the Mayor's Task Force (MTFE 1989b, May 2). Unlike the Edison meeting, however, this and all subsequent Task Force meetings went unreported in the *Chicago Tribune*. On one side were people who wanted the City to explore a wide range of options.[58] Collectively they attacked Edison for blocking the use of energy-efficient technologies, for corrupting the political process, for charging excessively high rates and imposing a regressive rate structure, for contributing to the decline of the City's public school system,[59] for failing to be honest and open about its program for cleaning up toxic chemicals, and for holding neighborhood meetings designed to "confuse" the public with "misinformation about a city takeover." Frank Rosen of the Labor Coalition and Sara Jane Knoy of National People's Action cut to the heart of the matter. Referring to the presence of CACI's president at the Task Force's meeting, Rosen passionately condemned the utility for corrupting the political process:

> Edison . . . is using the Chamber of Commerce from the inside to take away the only weapon you have when you want to talk about hard-nosed negotiations, because hard-nosed negotiations means you have an alternative. . . . I think if there is not a viable option to Commonwealth Edison, Commonwealth will do what it's been doing all along, which is talk, talk, talk, until

it's too late, then to use its friends in the courts and in the Congress and everywhere else where it buys influence and will again take us down the garden path to higher and higher rates.

And Knoy summarized the consumer groups' perspective: "It's Com Ed that should be asked to stop bashing the City of Chicago. This is not a personal thing; it's strictly business. All we want is the best and most economical alternative for Chicago, and we wonder why we're being criticized for that."

Also speaking to the Task Force on that day were three supporters of Commonwealth Edison and two other speakers who took a more neutral stance. Spokesmen for the CACI, the Committee for Sane Electric Power, and the Council on Religious Action insisted that Chicago business people feared a City takeover of Edison, that there was good reason to fear "ward committeemen or aldermen deciding who will throw the switch at nuclear power plants," and that Com Ed and its chair had made many contributions to the community. A spokesman for the Industrial Energy Consumers Association expressed concerns about maintaining reliable electric service and stable prices, and a researcher from the Heartland Institute (a conservative, not-for-profit research institute) praised the virtues of deregulation and increased competition.

A few weeks later, Charles Komanoff and Ralph Cavanagh from the KEA team briefed the Task Force on their findings up to that point (MTFE 1989b, June 28). Cavanagh outlined a scenario in which "the city would grow in its residential and commercial sectors, as it now expects to, for 30 years and yet at the end of that time end up using, for those two sectors, a third less electricity than current consumption levels." What is more, he said, those energy efficiency improvements would cost less than it costs Edison simply to operate its coal plants! The Task Force was obviously attracted to Cavanagh's scenario. More important, they were attracted to a "collaborative model" that he proposed: the City would work with Edison to change Edison's incentive structure so that the utility's stockholders would be able to profit from energy conservation investments.

By early August, the Task Force had made considerable progress toward developing an internal consensus about what the City should do. But it still did not know whether the Mayor and his aides would pay any attention to them or their recommendations. Further, the new administration had put a tight lid on information coming out of the Planning Department. Though KEA had submitted several draft reports and memos to the Department, none of them had been released to the Task Force or to the public. What is more, the administration had also stalled plans for Calmar Communications to conduct a public information campaign. Com Ed was reportedly "breathing a little easier with Daley in City Hall" (Dold 1989b).

Breathing easier, but still not breathing easy. Though Mayor Daley was not being very open about his intentions, he still was pursuing the possibility of terminating Edison's franchise. By late November it looked as though he would ask the City Council to notify Edison (by December 31, 1989) of the City's intent to terminate the franchise and to acquire utility property under the favorable terms specified in the franchise (Dold 1989c). Alderman Keith Caldwell, chair of the City Council's Energy, Environmental Protection, and Public Utilities Committee, indicated that he would hold public hearings in early 1990 (O'Connell 1990).

In late November, the Mayor's Task Force submitted its report to Mayor Daley (MTFE 1989a). The twenty-page report urged the Mayor to issue both notices in order to preserve the City's options in negotiations, and it urged the Mayor to give serious consideration to the Chicago Energy Commission's innovative purchase/lease-back proposal. In addition to emphasizing the importance of the City positioning itself to acquire Edison's facilities, the Task Force also stressed the importance of energy efficiency, a short-term franchise (if renewed), competition in supply (including reasonable wheeling charges), special attention to low-income residents, and local fabrication of energy-efficient products. "An entirely new arrangement for electricity supply is," it said,

> of paramount importance to Chicago in 1990. The nearly $1.8 billion that Chicago's residents, businesses, city governments, and schools pay every year for electricity is unacceptably high. It directly affects jobs and produces an economic burden felt by all, especially those least able to afford it (1) . . . The expiration of Chicago's present arrangement for the supply of electric energy presents a rare opportunity to establish new electric energy policies and practices that would make a significant contribution to the well-being of the City, its citizens and businesses. The Task Force encourages the Mayor and the City Council to seize the opportunity being presented (16).

Mayor Daley finally responded publicly to the Task Force's report by saying that he had an "open mind" about the possibility of City acquisition coupled with professional management (Dold 1989d). But by this time, the proponents of exploring options were deeply frustrated. They were underfunded, had no strong political allies, and were hindered by Daley's "close to the vest" approach to decision making, according to Moberg (1989). They feared that "the decisions will be made in closed-door meetings, with little public involvement and heavy-handed influence by Com Ed, just as there was with Rostenkowski's legislation." The new administration had canceled all of the Task Force's plans for public participation and education and downplayed any news about the Task Force or the franchise issue. City staff close to the issue were also reportedly frustrated and demoralized, not knowing where the administration was headed. Moberg concluded that Daley, who would be running for Mayor again in early 1991, would not attempt a major confrontation with Com Ed unless massive public pressure forced him to do so. And that pressure seemed unlikely to appear; neither CUB nor the Illinois Public Action Council (IPAC) had made the franchise a major issue, the CEOC was not advocating any particular alternative, and no strong political leader had emerged to galvanize the community groups' cause. "Our assumption is when we lost Harold [Washington], we lost the battle," said John Cameron on IPAC, "unless we can find some new horses with credibility." Com Ed, on the other hand, was wealthy, confident, and well-connected, and it had successfully made it appear that the City was on trial, not Com Ed. Moberg argued that

> The key to getting the best deal is the city's right . . . to buy the part of Commonwealth Edison that serves Chicago. . . . That is why Commonwealth Edison is doing everything possible to undermine that right (Rostenkowski's legislation for example) and to block serious discussion of acquistion with threats, scare tactics, ridicule, and a massive propaganda campaign against the "extremists" who promote its use.

On December 13 the *Chicago Tribune* praised the Mayor for keeping his options open and negotiating from a position of strength, but it urged him not to "dangle" acquisition, not even as a "bargaining chip." The idea "terrifies" most Chicagoans, the editorial said. Further, there is not enough time to carry out the acquisition, and it would simply result in more bureaucracy and "politicalization." The Mayor, said the *Tribune,* should focus on ways to improve customer service and to promote conservation and independent power delivered over Com Ed's lines.

Finally, on December 28, the City Council and the Mayor notified Com Ed that the City was terminating the franchise and would seek to purchase the utility's facilities. This did not mean, of course, that the City would actually acquire a portion of Edison's facilities. Nor did it mean that the City would follow the Chicago Energy Commission's sale/lease-back strategy. Rather, it simply meant that the City was in a position to take those actions if it wanted to. Conversely, it could negotiate a new, but scarcely modified, franchise with Edison. It all depended on the Mayor and the way he played his cards.

The Mayor's Hand Turns up Empty

Early in 1990, Mayor Daley formed a new Electricity Working Group to help him renegotiate the franchise. His manner of doing so implied dissatisfaction with the previous Energy Task Force, with Charles Williams and his energy planning staff, and with the City's technical consultants. Robert Helman, manager of a major local law firm, was asked to head the group and to lead the City's renegotiation efforts. KEA was asked to finish its final report, but the Working Group hired Portland Energy Conservation Group and another modeling firm as new technical advisors.[60] It was not clear whether the City would ever release any of KEA's reports.

From January through March, Alderman Caldwell's Energy Committee held its promised public hearings on the franchise issue at various locations throughout the city (Energy, Environmental Protection, and Public Utilities Committee 1990). Approximately 125 people spoke at those hearings. Many of them were former Edison employees and others whose economic livelihood was tied to Edison. They vigorously complained about the quality of existing City services and about the idea of a City takeover of Edison's facilities. Aldermen on the committee repeatedly stressed that City employees would not operate any new system, and they often emphasized the importance of negotiating from strength. Responding to CACI's recommendation that the City exclude the possibility of a purchase, for example, Alderman Caldwell said that would be like "swimming the English Channel in a cement overcoat" (Energy Committee, March 28, 81). Consumer and community groups were also well represented at these hearings. Agreeing about the importance of negotiating from strength, they repeatedly complained about the lack of publicity and attention that the City and the major news media had given the issue. Many of them also wanted to know how the City had spent the $400,000 that had originally been budgeted for the exploration back in 1987. It is worth noting that the *Tribune* continued its feeble coverage of the franchise issue by not reporting any of these hearings.

With but one notable exception, the hearings contributed little to the City's exploration of options. In the middle of the last hearing, Robert Helman outlined the process that the City would follow in negotiating with Edison. There were, he said, three options:

1. a purchase with immediate lease-back to Edison,
2. a purchase with an operating contract initially to Edison but later put out for bid to other potential competitors, or
3. a renegotiation of the franchise on the basis of five principles: rate reduction or containment, energy conservation, economic development, competition (through cogeneration), and environmental protection.

Two key issues would have to be resolved in any acquisition scenario, he said: (1) what facilities the City was legally authorized to buy and (2) how much the City would have to pay for those facilities. The key point was that, regardless of what the existing franchise agreement said, the City would probably not be able to buy Edison's facilities at "original cost." Either negotiations would result in a higher price, or else protracted litigation (which would take time and introduce risk) would drive the effective price up substantially. Any pursuit of the renegotiation option would be done in terms of a "compact" that would include a franchise of relatively short duration, retention of the acquisition option, and a way to measure Edison's progress toward complying with the five principles. "[T]hat is the thought process that we are following," Helman said. "There is no secret about it. There are no hidden arrangements about it. . . . Obviously [however], the negotiations . . . cannot be conducted in a public forum" (March 28, p. 114).

Helman's statement came on the last day of hearings and led many speakers to wonder why the hearings had been held prior to telling people what alternatives the City was seriously considering. Lew Kreinberg of the Center for Neighborhood Technology (CNT), for example, complained that if Helman's statement had been made at the start, then people would not be "just talking off the top of our heads or talking from public advertisement that's been carried on by Commonwealth Edison and we could really frame an intelligent conversation" (March 28, p. 147). Helman's statement also caused some to wonder who would benefit from his counsel not to conduct negotiations in public. Comparing the negotiation process to a poker game, Frank Rosen of the Labor Coalition argued that

> there aren't any secrets in negotiations. Now what cards you play at a given moment that's one thing. There aren't any secrets. If anybody thinks the working group, or any other, that Edison doesn't know everything that's going on in those meetings, they're crazy. . . . [T]he only secrets at this point are being kept from the City of Chicago, and I think that's wrong. (March 28, 232)

But Alderman Cullerton, who appeared to support Edison, responded: "I think you are wrong when you say it's just a question of when you play the cards. . . . It's very difficult to play the cards in any procedure, if you are dealing with open cards face up on the table. Also there's sometimes where you want to keep your cards—keep a card in the hold" (March 28, 238–39).

By May 1990, the negotiations between Edison and the Daley administration were well underway. Edison initially proposed to extend the franchise another forty-two years, reduce the franchise fee from 4 percent to 0.5 percent, and eliminate the buyout option contained in the existing franchise agreement. It also cited numerous court decisions that would, in its judgment, force the City to buy all of Edison's facilities or none of them. Conversely, Edison offered to consider lower rates for low-income and fixed income customers who use small amounts of electricity, to consider new conservation and energy efficiency measures, and to consider a few other minor items, including planting trees along its rights-of-way. "Everything is negotiable," said Vice President Petkus; "We have not closed any doors" (Dold 1990).

For the City's part, Mayor Daley announced that any new franchise would have to include a better deal for Chicago consumers, and—with his hand strengthened by Edison's April 1990 application to the ICC for an 18 percent rate increase and by information provided in the KEA report[61] (which the City had finally released to the public)—indicated that he was seriously considering taking over some of Edison's facilities to save Chicagoans money on their electric bills.[62] However, when asked whether he was serious about having the city operate nuclear power plants and other electric facilities, Daley laughed "No way. That would be a national disaster. City government should not run nuclear plants. One little mistake and we'd be done. It would be 'lights out'" (Davis 1990a).

"Lights out." An ironically prophetic phrase. Beginning in late July, as the negotiators appeared to be moving toward resolution, the largely black and Hispanic low-income West Side of Chicago was hit with a series of power outages. On the night of July 28, an estimated forty thousand homes and businesses lost power when a fire severely damaged one of Edison's switch houses. Blacked out for two to three days, and forced to bear the costs of spoiled food and random looting, West Side residents were reportedly angry, frustrated, and looking for someone to blame. Asked if the company would reimburse residents for losses due to the outage, an Edison spokesman said, "At this point, I don't know. We normally don't" (Reardon and Fountain 1990). Mayor Daley felt that the heat was on him, but—a *Tribune* editorial on August 1 said—"if he thinks that the heat's been on the last few days, he should imagine what it would have been like if the city owned the

FIG. 9.2
Danny Powell, Right, Grieving over Death of Three Relatives in a Fire Related to Edison Power Outage on July 28, 1990

electric system, as he has proposed in franchise negotiations with Commonwealth Edison."

Roughly one week later, the West Side was hit by a second blackout, this one lasting for about eight hours and affecting twenty-five thousand residents and businesses. Publicly referring to the blackouts as an "outrage," Mayor Daley proclaimed that Edison should reimburse its customers for their losses, and he appointed a city task force to investigate how utilities could increase the reliability of their service. Consumer groups argued that the blackouts were a consequence of the company's skewed investment priorities over the preceding ten years; i.e., its focus on building new plants. "Edison has diverted money to the construction program and used income to shore up dividend payments," said Susan Stewart of CUB; "That has meant postponing maintenance." Howard Learner of BPI agreed: "Something had to give, and over the last few years, Edison has shortchanged its maintenance and upgrade budgets for its transmission and distribution system" (Karwath and Barnum 1990).

Following so closely on the heels of the first blackout, the second outage stimulated a flurry of activity. The chair of the Illinois Commerce Commission, Terry Barnich, hired a consulting firm to investigate the blackouts. Consumer groups demanded that the company take responsibility for the shortages and compensate customers for losses without passing on the repair costs to customers. At a City Hall news conference on August 7, Mayor Daley said he was losing confidence in Edison and was keeping all options open in the franchise renegotiations. State and local politicians initiated a petition drive urging Governor Thompson and the General Assembly to require utility companies to waive monthly service charges when power is cut off for four or more hours. Edison's Chief Executive Officer, James O'Connor, responded to this activity by calling the two recent outages "unfortunate, isolated incidents" and by defending the company's overall performance and its recent rate increase requests. "We've been here over 100 years, and we've served the city pretty well," he said; "We hope to demonstrate that we are reliable and committed to improving our service" (Weinstein and Stein 1990).

These two electric power outages would appear to have strengthened the City's hand dramatically. However, the Mayor seems to have thought otherwise. In early September, he announced that the outages had short-circuited the negotiating proccss and indicated that he would offer the company a one-year extension of the current agreement. "We can't continue to negotiate until we have a clear picture of Edison's reliability," he said (Davis 1990b). Many observers suggested that the Mayor was less interested in strengthening his hand in the negotiations than he was in defusing the franchise negotiations as an issue in the mayoral primary scheduled for February 1991. Alderman Robert Shaw, for example, charged that "This is a sweetheart deal with Commonwealth Edison, pure and simple, because he doesn't want the residents of the West Side to see that he refuses to demand that Edison pay them for their

It's hard to imagine a clearer signal that Mayor Daley was not interested in revamping the institutional structure of electric power supply in Chicago or in negotiating a new franchise that would be much more favorable to the City and its residents. These outages could have acted as powerful tropes in the City's effort to persuade the public that a particular course of action was warranted. Rather than using them to mobilize support, however, the Mayor used them to justify inaction.

losses. . . . His opinion is that blacks have short memories, but he's wrong" (Davis 1990b). And a *Chicago Tribune* editorial on September 7 argued that if Daley signed a new deal with Edison, supporters of a city takeover might charge that he was a "craven tool of corporate interests"; if he tried to create a municipal power company, others would charge that he "had gone bananas."

On September 22, another blackout struck the West Side, this time lasting for about six hours and interrupting service to about fifteen thousand customers. Insofar as the outage was caused by an error made during routine maintenance and appeared to be a clear case of company negligence, Commonwealth Edison announced that it would reimburse customers for losses suffered from this third outage. One of Daley's press secretaries said that the mayor "thinks this third serious outage puts even more emphasis on the fact that the city needs more time to look into Commonwealth Edison's reliability before agreeing to a new franchise" (Rudd 1990).

Edison agreed to the proposed one-year extension just a few days after this third blackout. Negotiations continued, but in a context of ebbing passions.

At the beginning of 1991, the City initiated a $400,000 inquiry of Edison's transmission and distribution system. Critics of the utility hoped that the study would find that Edison's service was inadequately reliable, a conclusion that would increase the City's bargaining power. Citizens for Lower Electric Rates and Reliable Service, a new coalition of groups formed to influence the franchise negotiations, sought to further strengthen the City's hand by putting another referendum on the April 2 ballot. It asked: "Should the City of Chicago seek lower electric rates and more reliable electric service by acquiring Commonwealth Edison facilities within the city, with the facilities being operated by independent, professional managers for the benefit of Chicago ratepayers?" Edison considered it a loaded question, and dismissed the outcome in advance. Furthermore, Edison once again hired McKeon and Associates in February to ask eight hundred Chicagoans, "Who would you rather have providing your electrical service, the City of Chicago or Commonwealth Edison?" McKeon found that 67.5 percent said Edison, 19.7 percent said the City, and 12.8 percent did not know (Karwath 1991a).

Mayor Daley won the primary on February 26 by a wide margin—winning about 60 percent of the vote, to about 32 percent for Danny K. Davis (an African American candidate), and 5 percent for former mayor Jane Byrne—and then was reelected mayor in early April with 71 percent of the vote. He seemed to be in a very strong position to negotiate a new franchise agreement that would be quite favorable to the City.

In mid-June 1991, the City Council's Energy, Environmental Protection, and Public Utilities Committee began holding hearings on a new franchise. It sought to hear comments about two key options: (1) acquisition and lease-back of part of Edison's system and (2) a renegotiated agreement that would reduce rates for low-income households, promote energy efficiency and competition, enhance the reliability and safety of Edison's system, and have a much shorter duration than the existing forty-two-year franchise. Robert Helman told the committee that he was optimistic that an acceptable agreement could be reached. When doing so, he appeared to soft-pedal the acquisition option, saying that acquisition would probably end up in court, take several years to resolve, "and become a field day for lawyers" (Kass and Davis 1991a). He said his goal was to reduce monthly bills for consumers who use low to moderate amounts of electricity, to encourage the utility to create new jobs in Chicago, and to increase affirmative action opportunities for employees and minority contractors. Many aldermen seemed attracted to the emphasis on minority opportunity. Helman and the Energy Committee members appear, however, to have been caught off guard by the very hard line that the utility's chief negotiator, Susan Getzendanner, took during the hearing. According to her, the company wanted a nonnegotiable, forty-year extension on the franchise, and it thought that the 4 percent franchise fee should be eliminated. "In other words, you want the franchise for free?" asked Alderman John Buchanan. "I just want to show that we have some leverage power too," she said. Getzendanner also made it clear that, since the City would have to buy all of Edison's system, the utility's managers did not take the threat of a City takeover seriously. "We're talking a cost in the billions and billions of dollars," she said; "We have discounted a takeover as a threat. But we want to negotiate a fair agreement" (Kass and Davis 1991a). Mary O'Connell and Lew Kreinberg of CNT drew attention to Edison's aggressive efforts to sell all-electric construction for downtown high rises, to claims that Edison had blocked consumers' efforts to use alternative energy sources, to conservation issues, and to concerns about the reliability and safety of Edison's system (Maclean 1991). Alderman Bloom doubted that Daley had done enough to publicize the acquisition op-

Edison threatens to litigate for years, but it can afford to do that only if the ICC authorizes rate increases that enable Edison to litigate. Whether in their own self-interest or not, it is Edison's consumers who subsidize Edison management's ability to defeat efforts to alter the institutional structure.

tion to make it a viable bargaining lever, and Alderman Ed Smith condemned the hearings as a waste of time, saying that they were a political charade to justify an agreement that had already been reached. "I'm flabbergasted," he said; "I thought they [both City and Edison negotiators] would come in and give us honest information, but they were pompous, arrogant and vituperative. They have told us nothing. I'm convinced this deal is already etched in stone, and the poor people are going to suffer once again" (Kass and Davis 1991a).

Were Getzendanner's comments merely a pompous and arrogant bluff on Edison's part? Perhaps. But the next day's hearings gave Edison's arrogance much greater credibility. Expecting its $400,000 consultant to report that Edison had failed to maintain its distribution system adequately, the Daley administration appeared to be embarrassed when William Snowden of Failure Analysis Associates told the committee that Edison provided highly reliable service to Chicago residents. Snowden's dry, monotonic, jargon-filled lecture about advanced technologies dismayed the aldermen, and the content of his lecture irritated some administration officials. Recommending that Edison invest about $100 million in the distribution system, Snowden said "I tried to think up a story line to make this more interesting to you, but I couldn't. This is, after all, technical information." "This is unbelievable! This is worse than chemistry class!" yelled Alderman Smith; "He comes down here and gives a physics lecture, and he's got all the notes. What is this?" "We're in negotiations and this is a public event, and the focus wasn't what we wanted," said one official; "You're looking for 30 seconds of television time, a headline, and the phrase 'Edison is highly reliable' wasn't the one I wanted to see" (Kass 1991a). Some aldermen asserted that the City's negotiating position had been sharply undercut, and some consumer groups agreed. Indeed, several community group members left when Chairman Edwin Eisendrath (who had replaced Keith Caldwell as committee chair the previous year) told them there would not be time for them to testify. "It's more than an outrage; it's ridiculous that they'd keep us down here and then try to shut off public input," shouted Maureen Dolan of CEOC; "The city's already wimped out. All they want to do is bore everybody to death and cut a deal in secret." Samuel Mitchell of CACI had been allowed to speak. He told the committee that more than 28 percent

What a wonderful example of the mismatch between scientific planning and a political audience. Mr. Snowden couldn't think up "a story line" to make his presentation "more interesting," but everyone else was able to see that what made it interesting was its inconsistency with the Mayor's public story.

of city businesses would make plans to leave Chicago if the City came through on its threat to take over the distribution system (Kass 1991a).

"I'm no wimp," Mayor Daley declared the following day (Kass 1991b). Stung by Edison's presentation at the committee hearings, Robert Helman announced that the City was stepping up its efforts to acquire Edison's system. "I have never seen a more arrogant presentation," he said; "As a result, the city this week is stepping up sharply its putting together of a team" to possibly assume Edison operations after the lease expires in December. He said there was no way that the utility would get a forty-year extension or a cutback on its fee, and indicated that the two parties were drifting farther apart in negotiations. Edison's chief negotiator Susan Getzendanner countered by again dismissing the idea of a takeover as not being financially feasible (Davis 1991).

However much the Mayor and his chief negotiator insisted that Snowden had not been so tough on Edison as he should have been, and however much they proclaimed the seriousness of their threat to take over part of Edison's system, the damage had been done.[63] From that point on, the acquisition and lease-back option simply lost all credibility. With it went the City's ability to negotiate from a position of strength. Its best alternative to a negotiated settlement was to have Edison serve the City on the company's own terms.

By late June, it appeared that the City had been outwitted and outflanked by Edison. Embarrassed by political cartoons that made fun of his proclamation that he was not a wimp, Mayor Daley held a series of emergency strategy sessions with his aides, seeking to improve his bargaining position with the utility and wanting to portray his administration as being tougher. "All this stuff, it's just part of the negotiations," Daley said; "you just work it out; that's all." But aides worried about the public's perception of the administration and tried to diffuse responsibility: "We're doing what we can," one aide said; "The ICC sets the rates; they're ultimately responsible for reliability. Chicago's govenment can't step outside its framework. But we're doing all we can" (Kass 1991c).

Chicago's government, the mayor says, cannot step outside the ICC's framework. Odd. The whole point had initially been to explore the possibility of radically altering the institution of regulated natural monopoly.

In mid-August, Robert Helman publicly revealed that he had asked financial and engineering firms to draft a plan to take over Edison's facilities and that he was searching for a project manager to run the acquisition of Edison. He was doing so, Helman said, because "negotiations are not going well at all. . . . They're not willing to come forward on the sticking points in the talks" (Kass 1991d). In his view, Edison had balked on issues of price relief for low-income customers and providing an adequate energy conservation program, a system to measure reliability assessments of Edison's service in the city, and a stepped-up affirmative action program. Daley indicated that he was denying Edison's request to run a power transmission line above ground through the south side of town.[64]

The Mayor continued to state his support for the buyout option, but his words rang hollow. The City could do little more than seek marginally better terms, whereas Edison had one more card to play. In August, Getzendanner warned the City that Edison would stop paying the existing franchise fee after December 31, 1991, if no new agreement had been reached by that time.

The cards had been played and the Mayor's hand had, whether intentionally or not, come up empty. On October 22, the Mayor announced an agreement that would allow Edison to provide the city with power for another twenty-nine years. In return, Edison agreed to take several specific actions: it would extend rate reductions to elderly and low-income residents; it would spend $1 billion over the next ten years to improve its transmission and distribution system in the city; it would continue to pay the 4 percent franchise fee and to make its payments monthly rather than annually; it would not charge customers their monthly service charge whenever they suffered power outages exceeding twelve hours; and it would bury underground its new transmission line serving downtown Chicago. The mayor insisted that the City simply had been in no position to exercise the takeover clause in the expired contract. "Acquisition would trigger a very costly and bitter legal battle," he said, "something neither the city nor Edison is eager to undertake" (Kass and Davis 1991b). It was also clear that the Mayor did not want to lose the franchise fee during a time of significant budget deficit. Robert Helman sought to justify the pact by praising Edison's performance, and the Mayor indicated that rate decreases were the ICC's province, not the City's. Consumer advocates denounced the agreement, calling it "29 years more servitude to Commonwealth Edison" (Kass and Davis 1991b). The general impression was that the new deal looked much like the old one.[65]

The City Council Worries about Affirmative Action, Then Approves the New Franchise

Though the Mayor's hand had turned up empty, the game did not end for another two months. The new franchise agreement had to be approved by the Energy, Environmental Protection and Public Utilities Committee and then by the full City Council. The Energy Committee was scheduled to vote on the proposed franchise on November 20, and Chairman Eisendrath expressed confidence that—despite the opposition—he had the votes needed to approve it.[66] Some of those opponents charged that Edison and the proposed franchise were both racist. The Reverend Don Benedict, director of Clergy and Laity Concerned, was one. In his view, the franchise was racist because it did not protect low-income consumers (most of whom are minorities) from rate increases, and Edison was part of a racist power structure. He and other Edison opponents wanted the proposed agreement to be altered so as to reduce the cost to low- and moderate-income users of the first 400 kWh billed each month, to institute a conservation program of $40 million each year, and to limit the term of the franchise to five years (Kass 1991d). Other opponents, most notably Maureen Dolan of CEOC, argued not that the franchise was racist but that it was simply a bad deal for Chicagoans. In her view, the City Council should drastically amend the proposed franchise in five ways:

1. Its term should be reduced; twenty-nine years is too long.
2. Conservation should be its heart; Edison should be required to invest $200 million per year, or 2 percent of its revenues, rather than the $25 million proposed.
3. Bills for low-income consumers should be reduced more.
4. Safety and reliability guarantees should be delineated more clearly.
5. Enforcement and penalties for noncompliance should be specified.

Dolan acknowledged that the City might risk losing the franchise fee if the Council inserted these changes into the franchise agreement, but she argued that the City could always charge Edison rent for using City-owned rights-of-way for transmission lines (Dolan 1991). Despite these charges, the Energy Committee narrowly approved the proposed franchise.

A few days later, the full Council approved the proposed franchise (29–18), but only after incorporating an affirmative action amendment that leaders on both sides agreed effectively killed the agreement. The amendment, introduced by Alderman Bloom, would require the utility to award at least 25 percent of its purchasing contracts to minority-owned firms and 5 percent to female-owned firms. The defeat for Daley was reportedly the result of confusion, lack of direction, and bad planning on the part of his supporters and staff. At least three aldermen who were considered part of Daley's camp supported the amendment, one of whom said afterwards that she had been confused by the vote; another said he wanted to follow his conscience rather than the Mayor. Eisendrath supported the amendment, arguing that it was a minor change that wouldn't affect ultimate approval. (Davis and Kass 1991). A *Chicago Tribune* editorial on November 30 called the amendment "a short-sighted, ill-conceived, bone-headed move by the council, which has no business placing such conditions on a private company." Most of Edison's contract work goes to large firms, it said, that have no minority-owned competitors. Edison's opponents will find, it said, that Edison holds all the cards. Absent any agreement, Edison will continue providing electricity and collecting from its customers, but it will stop paying the franchise fee (which the city needs so desperately in a budget crunch). The Mayor tried to shrug off the council's action, saying he was in favor of affirmative action; "we'll have to work with them and see what happens," he said. An Edison spokesman called the amendment a "deal killer . . . a clear-cut intrusion by the city into how we do business" (Davis and Kass 1991).

On December 5, Edison announced that it would not accept the amended franchise agreement but indicated that it would agree to an amendment that put the the purchases of goods and services in terms of goals, not strict ratios. Edison also announced that it would pay its franchise fee on December 31, 1991, as required, but that it would withhold payments next year. It expected lengthy court battles over the fee and over city contingency plans to recoup the $70 million by charging the company for using city rights-of-way. A Daley spokeswoman said loss of the fee would hurt the budget and city services (especially police and fire, which comprised 73 percent of the City's expenses).

Edison's action and threat immediately set off a wave of finger-pointing and blaming among the aldermen. Alderman Edward Burke warned of financial chaos, arguing that the City would be unable to sell its tax anticipation notes, that its bond ratings would "be rated at junk bond levels," and that a City effort to charge for the use of rights-of-way would be tied up in the courts for years. Another alderman said that loss of the $70 million would cause thousands of layoffs in the police and fire departments. Alderman Bloom responded that the utility was bluffing and that Daley should stand firm. "If Daley had the backbone," he said, "Chicago could have had a champion on this issue instead of a wimp" (Kass 1991f).

A few days later, Daley's administration informed municipal union leaders that Daley would have to lay off nineteen hundred city workers to make up for the lost $70 million franchise fee. Alderman William Beavers, who supported the amendment, responded harshly: "I'm sick of Edison. They're trying to use everything they can. They've got kings showing, but I've got aces. I've got 26 votes on this thing. Affirmative action stays in the contract" (Kass 1991g). Union leaders (notably those from the Fraternal Order of Police, the Chicago Federation of Labor, the Chicago Firefighters Union Local 2, and Laborers Local 1001) said they were taking the threat of lost jobs seriously. They lobbied the aldermen hard. The Council was to vote the next day, and the vote was expected to be close. Alderman Bloom noted that Daley asked the union leaders to support him on this issue just as the City was renegotiating union contracts. "What kind of deal is Daley cutting with the unions?" Bloom asked; "What kind of chips is he calling in now?" (Kass 1991g). Apparently the Mayor was prepared to cast the tie-breaking vote against the amendment in the event of a 25–25 tie. Aldermen were reportedly offering wish lists to the Mayor's aides.

On December 11, 1991, the City Council gave its final approval to the new franchise by a vote of 26–23. Alderman Luis Gutierrez voted against the amendment, much to the anger of his African American and Hispanic colleagues. Gutierrez denied being pressured by Daley, but others said Daley had promised to help Gutierrez's bid to be elected to Congress. Aldermen shouted and denounced each other during the five-and-a-half-hour session. The Mayor's supporters passed a watered-down amendment that asked the utility to try to reach the percentages. "We were compared to David Duke, blamed for the Chernobyl accident, compared to George Wallace," said a utility spokesman; "We were blamed for causing Pearl Harbor; we're carrying all the sins of the world on our shoulders. So you follow the rhetoric, and it's another day at the City Council" (Kass 1991h).

FIG. 9.3 **Mayor Daley (left) Meeting with Alderman Luis Gutierrez on December 11, 1991**

After all the brouhaha of earlier years, after all the threats and bombast, after all the studies and counterstudies, it turned out that Edison held all the high cards.

The Discovery at the End of the City's Exploration: A Poverty of Imagination

At the start of this chapter, I posed three questions about institutional change. Why, I first asked, were the citizens of Chicago no longer content to work within the rules of the game and the given institutional arrangement, and why did they seek to change that arrangement instead? I would like to offer an answer that has economic, political, and technological dimensions. *Economically,* many citizens observed that Commonwealth Edison had initiated (with the ICC's approval) a massive nuclear power construction program that caused electric power rates to rise dramatically and in turn exacerbated the economic hardship already being experienced by the City of Chicago and its residents. They also observed that Com Ed and the ICC would not curtail that building program in spite of major reductions in the rate of growth in demand and the resulting increases in unused or excess capacity.

Technologically, many citizens observed that Edison's nuclear construction program locked northern Illinois into a large-scale and potentially hazardous technology just when advances in electric power conservation and alternative energy sources were creating an opportunity to base the electric power system (at the margin at least) on small-scale, diverse, and more efficient technologies. As public awareness of the technological alternatives increased, many people concluded that Edison's choice of technology was based less on a desire to provide power efficiently and reliably and more on a desire to maintain ownership and control over the assets involved in producing and distributing electricity; many concluded that Edison simply wanted to earn larger profits on an expanded rate base. Conservation, in contrast, would have met a portion of people's energy needs through decentralized decisions and small-scale customer investments that would have reduced their dependency on Com Ed. Technological change helped to call into question the prevailing folk view rationalizing the institutional arrangement that left the public totally dependent on a regulated "natural" monopoly.

Politically, many citizens observed that Edison's return on investment was rising dramatically just as its rates were becoming more burdensome, particularly to Chicago's low-income population. They concluded that the rules of the game were biased in favor of Commonwealth Edison and that those rules and the institution that fostered them needed to be changed. In particular, the rules of rate making under regulated natural monopoly granted significant advantages to the utility: Edison could use its considerable resources and permanent staff of attorneys, accountants, and public-relations experts to mount extensive and effective advertising campaigns, to garner political favors at all levels of government, and to sustain long and complicated rate cases before the ICC. To play the part of adversary in the rate proceedings game required community groups to make a huge and continuing commitment of time and money. These groups began to feel that this system was not capable of responding to their interests, given the high stakes involved and Edison's enormous investment in the status quo. The desire on the part of some citizens to change the rules of the game, in turn, fit nicely into Mayor Washington's progressive political and economic reform agenda.

So, many citizens of Chicago sought to change the institutional arrangement that authorized a regulated natural monopoly to own and manage the city's electric power system. They sought consideration of alternatives to this private monopoly, ranging from public ownership to a system of extensive competition. To what extent were those citizens successful in their efforts to change that structure, and what factors appeared to facilitate or obstruct their efforts? In the end, they did not succeed; indeed, the City's exploration of options and subsequent effort to negotiate a new franchise ultimately reproduced the existing franchise, with a slightly kinder and gentler face. We can say, however, that those citizens did succeed in putting the idea of institutional change on the City's political agenda, that they diligently kept that issue alive in mayoral campaigns, and that they were instrumental in causing the Mayor and the City Council to terminate the existing franchise and (though with uncertain commitment and ultimately with little success) to negotiate a new franchise agreement more favorable to the City.

To achieve this limited success, the community activists and City planners who supported the City's exploration of alternatives needed to go beyond commissioning technical studies by consultants and debating alternatives in open public hearings. They needed to "translate" the technical components of the various electric power options into activities that local business groups could recognize and to "enroll" those businesses and groups in the exploration by advancing their interests (see Latour 1987). To accomplish these tasks of translation and enrollment, the City's planners and citizen groups needed to network, negotiate, create coalitions, and seek to build a consensus around the options exposed and tested in technical research (see Benveniste 1989). These processes of translation, enrollment, networking, and coalition building seem to have been accomplished during the Washington and Sawyer administrations, largely through the task forces of 1985 to 1989 and the City energy planner's effort to actively mediate the process. The exploration of alternatives was part of Washington's political agenda, and the coalition linked to that exploration was part of his political constituency.

Largely as a result of these efforts at networking, negotiating, and coalition building, the City's planners and citizen advocates ("the progressive coalition") were—in spite of the power that accrues to a regulated natural monopoly—also able to induce the Mayor and the City Council to terminate the franchise with Commonwealth Edison. They were not, however, able to persuade the Mayor and the Council to restructure the institution or to negotiate a franchise that altered that institution significantly.

I want to suggest four reasons why the progressive coalition's efforts to change the institutional structure did not succeed. Surely one important factor was *the death of Harold Washington* and the change in audience that resulted from his death. Reelected in late 1987 with adequate support in the City Council, Washington was well-positioned to pursue whatever institutional option his advisors recommended to him. His death *led to fragmentation of the progressive coalition itself,* to vacillation on the part of his immediate successor (Eugene Sawyer), and to the 1989 election of Richard Daley. Daley proved to be a mayor very different from Washington. Rather than pursue a progressive reform agenda, Daley seemed much more inclined to support or accommodate the city's traditional growth coalition. Under him, the exploration of options seemed constrained by an overarching desire not to offend that coalition's corporate leaders. *Edison, in turn, placed itself in a very strong negotiating position.* Its political connections (especially through Congressman Rostenkowski) enabled it to weaken the City's ability to purchase Edison's system at a reasonable cost, and its financial resources enabled it to advertise its case widely and repeatedly and to threaten to litigate for years in the event of a City "takeover" of the electric power system. Who in Chicago (or points downwind) would want to have nuclear power plants run by precinct captains?

Thus the progressive coalition fragmented while Edison strengthened its position. This relative waning of the power of those who advocated institutional change was intensified by *the failure of electric-power rates to rise as dramatically as expected in 1985.* Instead of rising by 32 percent, Edison's rates had increased only by 8 percent as of the end of 1988, and even that increase had been overturned in the courts (see chapter 6). Edison did return to the ICC with a request for a 23 percent increase in mid-1990 (as will be shown in chapter 10). Though large enough to keep the heat on the new Mayor, this rate increase was not large enough to mobilize a powerful groundswell of opposition. And, unlike Mayor Washington, Mayor Daley displayed no interest in mobilizing that opposition.

At an even deeper level, however, the problem was that the planners and advocates who had been part of Mayor Washington's progressive coalition were unable to translate their disaffection with Com Ed into a persuasive account about how electric power could be supplied more fairly and efficiently in the future. They had no difficulty in conveying their disaffection with the institution of "regulated natural monopoly," but they had only limited success in extrapolating that story into the future (after all, Edison's rates did not increase as much as had been projected in the mid-1980s) or in helping Chicagoans to envision an alternative (and arguably better) way of providing electric power in the future.

What kind of story might the progressive planners and advocates have told, a story that would have enabled Chicagoans to understand the practical meaning of the progressive coalition's preferred alternative? I want to propose one possibility. In doing so, I am not claiming that it is the only story they could have told or even that it is one that they should have told. Rather, I present it more as a "thought experiment" that will help clarify the possibilities and consequences of treating planning as persuasive and constitutive storytelling about the future. So here is one future-oriented story that Chicago's progressive planners and advocates might have told, at least to one another and implicitly to others:

> In late 1989, a Task Force representing diverse interests in the city (including many members of the coalition) presented Mayor Daley with a report favoring fundamental restructuring of the relationship between the City and Commonwealth Edison. The Task Force labelled that new relationship a *community compact*. According to the compact, the franchise with Edison would be extended for another five years (subject to annual performance reviews), and the company would be required to invest $200 million annually into energy conservation measures. As its part of the compact, the City would agree to support legislation enabling Edison's shareholders to earn a reasonable return on energy conservation investments, and it would try to mediate a settlement between Edison and consumer groups over the company's contentious rate case. If the compact could not be negotiated, then the City would purchase part of Edison's system and lease it back to a private contractor (subject to public oversight).
>
> The Task Force's ability to forge this compact had been greatly facilitated by the work of a nationally respected group of experts (Komanoff Energy

Associates). KEA had provided the Task Force with a series of technical reports (including forecasts, models, and surveys) that provided good reasons to conclude that the proposed compact would produce substantially lower consumer bills as compared to continuation of the status quo. Commonwealth Edison challenged the merits of both the Task Force and KEA reports; however, Mayor Daley wisely decided to sponsor a series of community forums throughout the city. Those forums enabled the people of Chicago to understand and discuss the merits of both Edison's and the City's arguments and greatly strengthened the Mayor's hand in subsequent negotiations with Edison. In late 1991, the City and the company negotiated a new agreement that extended Edison's franchise for ten years (conditional upon biennial performance reviews) and that required the company to invest $150 million per year into energy conservation.

Though there were some stressful moments during the first year of the compact—successful mediation of Edison's last request for a rate increase was no simple matter—the company gradually began investing in local energy efficiency measures. Shareholders were pleased because they were able to earn a return on their investment in those measures, but customers were also pleased because their electric bills actually started to drop. What is more, a dozen or so small businesses began marketing energy conservation services and doing so quite successfully. Furthermore, community organizations like the Center for Neighborhood Technology (CNT) thrived as they built on their experience at delivering energy services to local businesses and neighborhoods. Rather than see their money flow out of the city to pay for imported energy, Chicagoans saw it being invested in local projects, thus contributing to a gradual improvement in the physical and social conditions of the city's older neighborhoods. Though Mayor Daley faced a tough fight for reelection in early 1995, most Chicagoans praised him for the agreement he had negotiated with Edison. The community compact had been good for the people of Chicago.

The first thing to notice about this story is that—unlike modernist planning would have it—it begins, as Toulmin (1990, 179) puts it, "*from where we are, at the time we are there.*" Furthermore, it emplots the flow of future action. In this case, the plot has what many readers might consider a utopian thrust to it: the plot presumes an ability and willingness on the part of Edison and the progressive planners and advocates to engage in meaningful negotiations with one another. Considering the ways in which each party had characterized the other in the past, this move to meaningful negotiation would have been an extraordinary turn indeed. After all, Edison was (in the eyes of the citizen groups) evil incarnate, and the citizen groups were (in the eyes of Edison's managers) ignorant fools who were trying to get rid of the company. Such a turn would have been possible only if each party had been willing to follow Hunter's (1991) advice: (1) to recognize the "sacred" within the other's moral community and (2) to recognize the inherent weaknesses in each community's own commitments.

Another key point to notice about my hypothetical story is that it reconstitutes the character of Mayor Daley. He continues to be a politician who wants to be reelected, but he is more willing to involve the general public in the negotiation process and to use the resulting public support as a bargaining lever in negotiations with Edison. This too might strike some readers as a rather utopian dream. And it is possible that Mayor Daley was irretrievably committed to Com Ed and its allies. But he was the Mayor, and it was he that would have to be persuaded. So a persuasive story about Chicago's electric power future would have to take him (and his objectives) into account.

The third point to notice about my hypothetical narrative concerns its setting. It is local: the city of Chicago in the early 1990s. This particular city is deeply divided racially and economically and has a very long history of rampant political corruption. To be persuasive to Chicagoans, the progressives' story would have to take those two key aspects of the locality into account. It could not, as Edison's successful "precinct captains" campaign persuasively demonstrated, conclude with "municipalization" of the company's facilities in Chicago, and it would have to improve the economic fortunes of the city's businesses and people, particularly those who are African American, Hispanic, or of lower income.

The narrative is also written from a believable point of view. Once again, this presumes a shift on the part of the coalition. Its spearhead group on the franchise issue (the CEOC) had been portraying itself as a coalition of nonpartisan, citizen-based organizations that wanted to "educate" the public. Despite this self-characterization, their rhetoric had been quite one-sided, referring to Edison as "peddlers of panic," a "utility octopus" that was trying to maintain its "stranglehold" over the city. In my hypothetical narrative, the progressives back off from such inflammatory talk. They continue to advocate a major shift from the status quo, but they are very clear about their own normative stance when doing so. They stand for a fairer distribution of wealth, power, and income, and their story has them successfully pursuing that fairer distribution. Rather than focusing on whether Edison will continue to be the monopoly supplier of electric power, they concentrate on changing the conditions under which Edison supplies that power.

Lastly, the hypothetical narrative incorporates a few distinctive tropes, each of which makes the overall narrative more persuasive. The Task Force's *claim to be broadly representative* of the people of Chicago is one, and KEA's *technical arguments* in favor of the compact is another. And surely the series of *community forums* act as a powerful trope by providing Mayor Daley with the political support he needed to negotiate from a position of strength. One other trope worth noting, though surely there are others, is the phrase *community compact.* The public needed a simple phrase or slogan to help them understand the progressives' alternative. Prior to 1990, the closest the coalition had come to such a phrase was *municipalization.* A metaphor without a soul if ever there was one. "Community compact" conveyed the much needed sense that the progressives' alternative would end (or at least radically temper) years of costly conflict between Edison and its opponents and do so in a manner that would be good for Chicagoans.

Unfortunately, my story is purely hypothetical. The coalition's groups chose a different tack, choosing to attack Edison and the Mayor. Left to their own devices, Edison and the Mayor negotiated a franchise agreement quite different from the one contained in my hypothetical tale. Left out in the cold, but with their moral commitments intact, the progressive planners and advocates could only help consumer groups continue their bitter struggle with the hated "utility octopus."

Interlude

Attending an Annual Meeting of Edison Shareholders

Two days ago I was walking through an old wooded ravine on the east side of Iowa City, watching a hawk circle and screech high above me. Today I find myself striding toward the Chicago Hilton Hotel for the 1993 annual meeting of Edison shareholders. Accepting a flyer from the twenty or so demonstrators who picket the entrance, I enter the building and find my way to the Grand Ballroom. Embroidered with Empire-style gold laminations, and brightly lit by ten huge glass chandeliers that hang from the ceiling, this vast room is filled with fifteen hundred to two thousand shareholders. Politely quiet, they gaze expectantly toward the elevated platform that seats twenty-eight company directors and officials. Moving to a chair at the far front right of the audience, I hear Chairman James J. O'Connor elaborate on his theme that "1992 was a year of contrast and contradiction" and Samuel K. Skinner, the new president, forcefully stress how company officials need to "correct public misunderstandings" about Edison's rates and reliability. Chairman O'Connor then deftly fields questions from the floor. It quickly becomes clear that my fellow shareholders are motivated by one concern—to maximize their total return on investment—and that their dominant mood is one of anger toward CUB and the ICC for having caused their shares to erode in value. One by one, the twenty or so shareholders who speak complain that the company "seems to be the chief victim of CUB's propaganda" and that the ICC is "the worst regulatory agency in the country." They want to know "who's running this company," CUB or professional managers? And they urge Edison's managers to "take off your gloves and fight!" But a minority of shareholders turn the blame toward the company's managers. "I have no faith in this board," says one. Listening to them speak, I feel radically disconnected from the wooded ravines of eastern Iowa City and the vital social world of Chicago, and I am profoundly struck by the shareholders' lack of understanding (or even curiosity) about what motivates consumer and community groups to act in the political world. The meeting adjourned, I walk out the door to Grant Park. Feeling the wind in my hair, the sun on my skin, my heart passionately engaged with humanity and the natural world, I realize that (for me at least) few things can be more evil than to translate the rich vitality of life into the single-minded pursuit of maximum return on investment.

The rate case from hell.

— COM ED CHAIRMAN JAMES J. O'CONNOR

Frozen in a Passionate Embrace: Allocating Pain, Allocating Blame

The effort to persuade the people of Chicago to embrace a new institutional structure for supplying electric power failed. Instead the "abnormal discourse" of electric power planning in Chicago simply reconstituted (with very minor exceptions) the institution of franchise-limited regulated natural monopoly. We find ourselves, therefore, drawn back to the ICC's hearing rooms, the contentious counsels, and the commissioners' efforts to formulate a rate decision that would satisfy Commonwealth Edison, the consumer intervenors, and—most importantly—the Illinois courts.

Recall that 1989 had ended on an ominous note for Com Ed. The company's electric sales had grown by an anemic 0.9 percent, peak load had actually declined by 3.9 percent, the Supreme Court had just overturned the 1988 "settlement agreement" that was not a settlement agreement, and the City of Chicago had terminated its franchise with the company. The year had ended on a sour note for the Commission, too. It had just suffered its third "embarrassing" court defeat in the last two and a half years. In mid-1987, the Supreme Court had overturned the ICC's rate increase for Byron 1; in June 1989, it had overturned the Commission's decision on Edison's differential between summer and nonsummer rates; and in December of that year, it had tossed out the "settlement agreement."[67] Commissioner Romero said the Supreme Court's decision that the 1988 settlement agreement had been illegal was "a lesson for us. You can't do a backdoor deal and call it adequate" (Karwath 1990b). No more experiments for the ICC. Now it intended to adhere strictly to the Public Utilities Act.

More and more, the modernist expansion plan of the early 1970s seemed to be leading Edison into a corporate disaster. More and more, the modernist regulatory apparatus seemed to be leading the Illinois Commerce Commission into a deadly cross fire between consumer groups and the courts on one side and its corporate clientele on the other. With Edison and consumer groups frozen in a passionate embrace, and with the possibility of inventing an alternative in Chicago foreclosed for another twenty-nine years, the institution of regulated natural monopoly (and the modernist story that continually reconstituted it) seemed to teeter on the brink of collapse.

Thwarted Again: The Courts Reject Another Rate Decision

On April 12, 1990, just months after Chicago had terminated its franchise with the utility, Edison returned to the ICC, asking for a rate increase of 17.7 percent or $982 million. This application came while the Commission was still waiting for instructions from the Supreme Court about whether Edison had to issue refunds and lower rates to compensate for the increase that the Court had declared illegal in December 1989. Consumer groups said that they would oppose the request, and Howard Learner complained that Edison was complicating matters by filing a new request before the old one had been resolved. "It's a thinly disguised means of heading off the much-delayed refunds to which consumers are entitled," he said. A spokesman for Edison rejoined that the lack of rate increases to cover the cost of the new plants was hurting the company: "Our earnings per common share dropped more than 40 percent since 1987," he said, "and are below the dividend we're paying out (Karwath 1990c). And, as had been the case in 1986, the Edison rate increase became a part of gubernatorial politics. Long-time Governor Thompson had decided not to seek reelection, and the two candidates to replace him (Democratic Attorney General Hartigan and the Republican Jim Edgar) both criticized the rate request. Edgar claimed it was too high and Hartigan promised that his office would intervene to oppose the boost.

On the last day of May, the Illinois Supreme Court gave the Commission the instructions it was waiting for: it ordered Edison to refund the first step of the 1988 rate increase ($235 million) plus interest, bringing the total refund to perhaps as much as $375 million. The illegal increase would also have to be removed from customers' bills. Edison called the decision a "severe financial jolt"; its stock dropped $1.25 per share to $31.62. (Edison's earnings for the first quarter of 1990 had already dropped 42.6 percent over the previous year.) Edison officials pledged not to reduce their dividends, implying that the company would have to dip into cash reserves. "Long term, they will get a rate increase," one financial analyst said. "But the longer it takes, the worse it is for the company." Attorney General Hartigan and Mayor Daley hailed the Court's decision. The Court order gave the ICC eleven months to determine how much of a rate increase, if any, Edison should receive (Karwath and Arndt 1990).

On June 14, the ICC instructed its hearing examiner to draft an order instructing Edison to remove the first step of the 1989 rate increase and to refund all the money collected for the eighteen months it had been in effect. Edison argued that it should be allowed to continue billing customers at current levels, despite the Supreme Court's finding that the rate increase had been illegal. The company further argued that the future of the rate increase should depend on how the ICC responded to its April request for a 17.7 percent increase. Twelve consumer groups and government agencies quickly threatened court action if the ICC went along with Edison's request.

Two weeks later, the Commission ordered Edison to refund approximately $400 million to its customers, by far the largest refund in the state's history (Karwath 1990d). One day later, the Commission ordered Edison to roll back its rates to the levels that existed before January 1989, and it terminated other provisions of the 1988 order. Edison's request for a 17.7 percent increase had not included the rollback; so on July 20, the company announced that it was boosting its rate increase request from 17.7 to 23 percent (approximately $1.23 billion).

In late August, ICC staff argued that Edison should receive about half of its request (11.3 percent or $598 million). The staff also indicated that the ICC could apply a different accounting method that would result in a $782 million, 14.8 percent increase. Staff recommended a partial rate increase of 2.3 to 4.4 percent to take effect as early as January. Consumer groups doubted that any increase could be justified, whereas Edison said the staff's recommendation came up "far short." Staff advised that all of Byron 2 and Braidwood 1 and about 45 percent of Braidwood 2 were needed to meet demand and provide Edison with a 20 percent reserve margin. It also suggested that $1.4 billion of the $7.1 billion total cost of the three units resulted from delays attributable to Edison or its contractors (Karwath 1990e).

Consumer groups were quite skeptical that the Commission would produce an acceptable decision, having concluded months earlier that Governor Thompson's newest appointees to the Commission made it even more pro-utility than before. This too became part of the gubernatorial election. Just days before the November election, Neil Hartigan said that, if elected, he would fire Terry Barnich as chair of the

ICC. Charged with favoring utilities, Barnich retorted, "I don't know what he's talking about. I try to dispatch my duties as I see best under the law" (Anonymous 1990). Unfortunately for Hartigan, the Republican candidate Jim Edgar won. Edison's stock climbed 50 cents to $33.75 per share in response to the news.

Near the end of 1990 and the start of 1991, just as Com Ed and the City of Chicago were beginning to renegotiate the franchise agreement, two sets of hearing examiners proposed orders concerning both the remanded proceedings and Edison's request for a 23 percent increase. The proposed order in the rate proceedings provided for an increase of approximately $579 million in the company's revenues.[68] This recommended increase reflected the examiners' judgments about whether and how to include Byron 2 and Braidwood in the rate base. In the remanded proceedings, they concluded that approximately $730 million of the roughly $7 billion cost of Byron 2 and Braidwood had been "unreasonably" incurred and should be excluded from the rate base. In the rate proceedings, they concluded that all of Byron 2 and 29 percent of Braidwood 1 were used and useful but that none of Braidwood 2 was. They recommended that the company be allowed to earn a return of 11.1 percent on the used and useful portion of the reasonable construction costs of the units and a return of 5.3 percent on the remaining $3 billion. The examiners also recommended that the company be allowed to recover approximately $1.16 billion of carrying and depreciation charges that had been incurred after the in-service dates of the three units. This recovery would take place over a five-year period (Com Ed 1991). Consumer groups derided the proposal: Howard Learner of CUB complained that "Examiner Freund is whitewashing the independent auditors' findings that there was substantial waste during the construction of the Byron and Braidwood plants." Edison management appeared no happier; a spokesman referred to the examiners' proposal as "a punitive draft order" (Karwath 1990f).

On March 8, 1991, shortly before the Commission was scheduled to decide on Edison's request for a 23 percent rate increase, Governor Edgar announced that he would not decide whether to reappoint Barnich as chair or Calvin Manshio or Jerry Blakemore (whose terms had expired in January) as commissioners until after the Commission rendered its decision. He did not have to wait long, for the Commission decided the case on the same day, approving 6–1 an aggregate annual increase of approximately $750 million (ICC 1991; Karwath 1991b). The rate increase would be phased in over a four-year period in order to avoid rate shock. Rates would increase by 9 percent ($483 million) immediately. They would rise an additional $231 million in 1992 to

13.3 percent, and another $241 million in 1993 to 17.8 percent. But in 1994, they would decline by $205 million, resulting in a 14 percent increase over the rates of early 1991. The increase allowed Edison to recover approximately $1.73 billion of deferred carrying, depreciation, and decommissioning charges on the units. (This pertained to the gap between when the plants were put in service and when the rate increase was granted.) It also allowed the company to include in its rate base approximately $480 million of unaudited capital additions to electrical generating facilities. Conversely, the Commission found that approximately $734 million of costs had been "unreasonably" incurred and hence were not to be included in the rate base. The Commission also concluded that, contrary to its own staff's advice, all three units were "used and useful."

Not surprised by the general tenor of the decision, BPI, CUB, the Attorney General's office, and other consumer intervenors immediately sued, complaining that the plan gave Edison too much money for the plants, two of which they claimed were not needed to meet demand.[69] Responding to the consumer groups' suit, the Illinois Appellate Court in Chicago put off the rate hike until those groups could challenge its merits in the Appellate Court or in the Supreme Court. On March 14, the Supreme Court agreed to hear challenges to the rate increase; but on March 19, it authorized Edison to begin charging the first phase (while still indicating that it planned to assess the legality of the rate increases). If the Court overturned the ICC's decision, Edison would have to refund money to customers, along with 5 percent interest. Consumers might have responded to this portion of the Court's order with some skepticism. Despite orders, Edison had still not paid refunds on the 1985 rate hike, which had been overturned by the Court in 1987.[70]

The Commission's decision pleased Edison's supporters. The *Chicago Tribune,* for example, supported the increase while also arguing that the experiences of the past four years demonstrated that the Public Utilities Act needed to be amended. In an editorial on March 14, it argued that

> Edison needs the revenues to pay for the plants and to get on with properly maintaining its system and improving customer service. Illinois needs to get on with rewriting the rules of the game to minimize expensive, prolonged litigation and allow regulators to try nontraditional approaches to ratemaking that will create a stable environment for economic growth. . . . In the current morass, Illinois utilities are hesitant to make major investments because there's no guarantee they'll be able to charge rates that allow them to meet operating costs and repay lenders. . . . The General Assembly can help achieve that by

revising state utility law to allow negotiated rate settlements and other alternatives to today's cumbersome, costly process. Ways must be found to end the rancor and demagoguery, to curtail litigation and give Illinois progressive regulation that is fair to consumers and to utilities.

The Supreme Court heard oral arguments on April 25 at its Daley Center chambers in downtown Chicago. Consumer groups charged that the three plants would not be fully used and useful for years and that the Commission illegally gave Edison $1.7 billion to compensate it for three years that the plants were generating electricity without a rate increase. Edison responded that the plants are needed and are "used and useful under any test" (Karwath 1991c).

While they were appealing the Commission's decision, consumer groups were also trying to document their claim that the commissioners (Chairman Barnich in particular) were biased in favor of Edison. By November, they had produced some compelling documentation. They reported that Commission members had made 536 phone calls in 1990 to officials and representatives of regulated utilities, while making just four calls to consumer groups over the same time period. Nearly half the calls were made by Chairman Barnich, and—despite a state law that bars ICC members from talking to utilities about rate cases—many of them came at key times in electricity rate cases and during court and legislative review of telephone regulation. Barnich said that he never discussed rate cases during any of those phone calls. Rather he was just calling "three of my best friends" who just happened to work for or else be closely affiliated with Commonwealth Edison: Michael Hasten and Phil O'Connor, both of whom were former ICC chairmen, and James Fletcher, an ex-aide to former Governor Thompson (*Chicago Tribune* 1991c). The appearance of improper communications was difficult to deny.

Much to Edison's dismay and the *Tribune*'s chagrin, the Illinois Supreme Court once again found the Commission's decision defective. On December 16, 1991, it ordered the ICC to reconsider several aspects of its March 8 decision to grant Edison a rate increase phased in over four years. In its decision, the Court again faulted the Commission for not following its own rules in determining how certain costs were to be figured. The Court reversed and remanded the ICC's finding that all three units were used and useful, and it said that the Commission never formally determined that continued construction of Edison's last three nuclear plants had been "prudent." The Court instructed the Commission to make a proper determination of the prudence of the continued construction of the Braidwood units; and it ordered the Commission to reconsider whether the rate increase should have taken into account the $1.7 billion of deferred carrying, depreciation, and decommissioning costs. The Court did, however, uphold the Commission's finding that more than $700 million of the three units' construction costs had been unreasonably incurred. The Supreme Court did not order the Commission to roll back or refund the initial 9 percent increase (Illinois Supreme Court 1991).

The Court paid close attention to whether construction of the units had been *prudent,* whether their construction costs had been *reasonable,* and whether the units were *used and useful.* To determine the prudency of building the units, the Court had to construct a history of Edison's expansion plan and the ICC's response to it. That history drew attention to the Commission's granting of certificates of convenience and necessity in 1973 and 1974; to its October, 1980, finding that both ratepayers and Edison would benefit if the Byron and Braidwood units were completed in "as timely and economic [a] manner as possible"; to the Commission's 1982 order again directing Edison to complete the units in a timely and economic manner; to the investigation of whether Braidwood should be canceled; and to the Commission's 1986 order requiring Edison to show cause why Braidwood 2 should not be canceled. Agreeing with consumer intervenors, the Court concluded that the Commission had failed to determine the prudency of continued construction after 1982. The ICC did not order Braidwood 2 to be canceled; neither did it approve continued construction, as required by the Public Utilities Act. Accordingly the Court instructed the Commission to make a proper determination of the continued construction of the Braidwood units.

Notice the intimate connection between *prudency* and planning as persuasive storytelling about the future. The Commission was supposed to determine the prudency of continuing Braidwood 2's construction on the basis of *need,* which depended on forecasts of future demand, which in turn depended on the company's and the Commission's efforts to encourage energy conservation. Whether Edison's construction was prudent depended on which story, Edison's or the consumer groups', the Commission embraced.

The Court also assessed whether the Commission had properly determined what costs were reasonable. It first noted that, advised by the construction cost audits, the Commission determined that $734 million of the $7 billion had been unreasonable. Consumer intervenors contended that the Act required the Commission to consider whether Edison's actions "resulted in efficient, economical and timely construction" and were based on "knowledge and circumstances prevailing at the time." They argued that the Commission had failed to consider the first prong of this test. The Court found the Commission's interpretation of the Act persuasive, primarily because it would be unfair to conclude, "based on a reading of the entire Act, that the legislature intended . . . a utility to prove that its actions were reasonable using hindsight" (Illinois Supreme Court 1991, 1045).

The Court also examined the Commission's judgment about whether the units were used and useful. The Court noted that all parties agreed that pre-1986 used and useful standards apply because the units were under construction prior to the enactment of the 1986 Act. Therefore it asked whether the "needed and economic benefits test" set forth in the 1986 Act was an appropriate criterion to employ. The Commission had found that a prior Appellate Court decision had precluded the ICC from using that test and that, according to other criteria, the units were entirely used and useful. The Court disagreed with both the Commission and the Appellate Court: "Our reading of the statute yields a different result. We believe section 9-215 is intended to place limits on the Commission's discretion only for those plants not already under construction on January 1, 1986. With respect to plants already under construction before that date, section 9-215 places no limits on the Commission's discretion. Rather, the statute permits the Commission to use a wide range of tests, including the needed and economic benefits test. This interpretation is supported by a review of the legislative history of the amended Act" (1053). Concluding that the Commission had been unduly constrained by the Appellate Court's previous ruling, the Court remanded the issue to the Commission. It did not, however, order the "needed and economic benefits test." Whether that test would be used was left to the discretion of the Commission.

Sections 9-212 and 9-215 of the Public Utilities Act state that a facility "is used and useful only if, and only to the extent that, it is necessary to meet customer demand or economically beneficial in meeting such demand . . . The Commission shall have power to consider, on a case by case basis, the status of a utility's capacity and to determine whether or not such utility's capacity is in excess of that reasonably necessary to provide adequate and reliable electric service."

Altogether, then, the Court found that the Commission had failed to determine whether construction of Braidwood had been prudent, it agreed with the Commission's standard for determining the reasonableness of costs, and it found that the Commission had been unduly constrained in determining whether the last three units were used and useful. Once again, the Court had found a Commission decision seriously defective. "It truly undermines consumer confidence in the fairness of the ratemaking process," Howard Learner of CUB said, "when the ICC and Edison consistently try to circumvent the law and the Supreme Court has to keep reversing those decisions" (Karwath and Grady 1991).

One week later, various intervenors asked the Supreme Court to suspend the first step of the March 8 increase and to place the amounts previously collected pursuant to that order in escrow. On February 3, 1992, the Supreme Court ruled that the first step should remain in effect while the ICC tried again to decide how much Edison's customers owed. It also ruled that the second and third steps of the March 1991 rate hike should not take effect (Com Ed 1992).

So Edison, the consumer groups, and the Commission (whose chair was charged with bias in favor of the utility) had to begin the rate case again.

The "Rate Case from Hell" Just Keeps Going and Going

The Commission began reconsidering Edison's rate increase shortly after it learned of the Supreme Court's ruling. This time, however, a new group of commissioners would be deciding what to do. By early January 1992, Governor Edgar was trying to decide whether to replace Barnich, Manshio, and Blakemore, all of whom had voted for Edison's latest rate increase. Barnich simplified matters for the Governor when he resigned as chair, but not as a member of the Commission, on February 13. He spoke of growing tired of "ceaseless personal vilification" from CUB and its allies and of feeling that "the politicization of the regulatory atmosphere has largely stifled innovative and progressive approaches" to utility regulation. Blamed by Barnich, consumer groups were pleased with his decision to step down as chair (Karwath 1992a). Barnich said that he suffered "more personal harassment" a week later when the *Chicago Tribune* reported that he had made fifty-nine phone calls to Edison officials and consultants in the two and a half months prior to the March 8 rate decision and that he had called an Edison vice president, Edison's law firm, and one of Edison's consultants the very day that the Appellate Court reversed an ICC decision on July 15, 1991. Susan Stewart of CUB said that "it is an incredible pattern suggesting that there was improper communication," and she indicated that CUB would ask Barnich to excuse himself from current Edison rate cases (Karwath 1992b).

A few days later, Governor Edgar appointed Ellen Craig as interim chair. The only commissioner to vote against Edison's rate increase in March 1991, she had come to be thought of as the most pro-consumer voice on the current Commission. Several days later, in mid-March,

Reconstituting Consumers as "Innocent Little Girls"

One major factor contributing to Edison's "bad public image" was the quantity and rhetoric of its advertising. The Public Utilities Act allows electricity companies to charge consumers only for advertisements promoting safety or conservation. It requires that ads promoting the company's image (such as the "We're there when you need us" campaign described in chapter 7) be paid for by the shareholders. But Edison had been advertising heavily (budgeting $6.6 million in 1990), and the advertisements did more than promote safety or conservation. Given the way that the ICC arrived at rate decisions—focusing on the bottom line of how much rates would increase—the company had no difficulty passing the costs of these and other advertisements along to the customer. The advertisements in turn rhetorically constituted a particular kind of consumer. Here are two noteworthy examples.

Edison's "little bunny in the road" advertisement appeared in February and March 1990. I had hoped to reproduce that advertisement at this point in the text, but I have been told by Edison staff that the "littly bunny" does not fairly represent the comopany's general advertising policy at that time. I cannot, therefore, reproduce the ad here. Imagine a cute little bunny rabbit sitting on the white line of a road, looking toward the reader. Two headlights glare menacingly out of the darkness, rapidly approaching both the bunny and the reader, and thereby implying that the cute little bunny is on the verge of being flattened onto the asphalt. Directly below the bunny, in large print, are the words: "Some Utilities Are Only Now Seeing What We Saw Coming 20 Years Ago." Another 150 words of text in smaller font size elaborates on that theme. By means of this advertisement, Edison portrays itself as a company that is able to foresee the future demand for electricity and to identify and build power sources that meet that demand at a reasonable price. Whereas other parts of the country are being blindsided by demand increases that they did not foresee, much as the poor little bunny is about to be blindsided by the car, Edison is not—because it built "a source" that minimizes air pollution and conserves oil and gas resources. The "source" is, of course, Edison's final set of nuclear units. The company does not say that Edison had forecast a growth of 7 percent per year rather than the less than 3 percent that actually occurred. Nor does it say that the "source" cost $11 billion more than originally planned. *Who is the little bunny in the road, and who is driving the car?*

The "innocent little girl" (fig. 10.2) appeared at the same time. Edison again portrays itself as a company that plans for and meets the electrical needs of the people of northern Illinois. Edison will "assure her of a start in life without shortages, blackouts or rationing." And "if we achieve our goals, she may never give her

electrical service a second thought. Which is precisely what we continue to strive for." Just three months after Edison ran this ad, Chicago's West Side experienced a series of supply interruptions that blacked out thousands of homes and businesses for days. Yet the ad prefigures the type of citizen and consumer that Edison apparently wants to create: naive, unthinking, uncritical, trusting, totally dependent on Edison for vital services. ***Who is the little girl and why does Edison want her to remain so "innocent"?*** [71]

We've Already Given
More Thought To Her Electrical Service
Than She'll Give It In A Lifetime.
Twenty years ago, we were planning the nation's most modern and efficient electrical generating system to assure her of a start in life without shortages, blackouts or rationing. But she doesn't know that.
Now, with capacity sufficient for years to come, we've redoubled our efforts to upgrade and improve the transmission and distribution system that will carry her electricity more reliably than ever. She doesn't know that, either.
To provide these benefits at the greatest value, we're renewing our efforts toward Least Cost Planning. We're investigating new methods of conservation, as well as looking at alternative generating technologies which can help the existing supply of electricity go a lot further.
And though she doesn't realize it now, we're also planning new and improved customer services to better anticipate and respond to her needs.
The fact is, if we achieve our goals, she may never give her electrical service a second thought. Which is precisely what we continue to strive for.
Commonwealth Edison
We're There When You Need Us.

FIG. 10.2 **Edison's "Innocent Little Girl" Advertisement**

the Governor nominated two new people to replace Manshio and Blakemore and help ensure the "confidence of the public" in the ICC: Karl McDermott, currently director of the Center for Regulatory Studies in Springfield, and David Williams, Jr., a former public works chief and state Department of Transportation official (Karwath 1992c).

FIG. 10.1 **New ICC Chair Ellen Craig**

With two new members and a new chair, it seemed that the Commission would do almost anything to ensure that its decision would stand up in the courts. To do so, it would have to determine whether Braidwood 2's construction had been prudent and whether Edison's last three nuclear units were used and useful.

In early March, Edison argued that it should be allowed to recover approximately $500 million in depreciation and other costs accrued from 1987 through 1991 and that the Commission should set aside the rule requiring utilities to use financial data from a single test year. Citing Edison's effort as another example of the utility's "arrogance," consumer groups cried once again that Edison was trying to persuade the Commission to violate the Public Utilities Act. "This is another step in Edison's continual effort to sidestep the legal requirements designed to protect consumers," Learner said. And an attorney representing the Cook County State's Attorney said that "they seem to be operating on a different plane from everybody else" (Karwath 1992d).

In April 1992, an audit prepared by Richard Metzler and Associates for the ICC concluded that Edison had unnecessarily required its customers to pay for overpriced coal (Edison was paying three times the market price for low-sulfur coal from Montana and Wyoming). The audit also determined that the company had built nuclear generators it did not need. Even assuming a 20 percent reserve margin, it said, Edison "is in the enviable position of having approximately 2,000 megawatts of surplus capacity" and—by implication—does not need the two Braidwood units. Thirdly, the audit observed that Edison had developed a bad public image. The company, it wrote, is often perceived as one that "deals only with legislators, business leaders and other power brokers and ignores or bullies community groups and municipalities." Edison's president responded: "The auditors don't need to tell us that. It's quite clear that the general perception of the company is not what we'd like it to be." Lastly, the audit also recommended that Edison should expand its energy conservation programs and improve the maintenance of its nuclear plants. These reforms could, it concluded, save Edison's customers $157 million immediately and $193 million each year after that. Edison retorted that the auditors' suggestions could, at best, produce a one-time savings of $8.7 million and annual savings of $4.9 million. Edison told the Commission that it had already begun implementing some of the auditors' suggestions. The Commission had ordered the 1989–90 audit—which took a year to produce, at a cost of $2.6 million—to comply with the Public Utilities Act. Susan Stewart, executive director of CUB, said, "This will back up our argument in the rate case that we should not be billed for a plant before it is needed." But Edison's president retorted that the audit's finding that the utility had too much capacity was misleading. According to Edison's records, it had a reserve margin of only about eight percent on July 22, when the company reached its peak demand of 17,733 MW (Karwath 1992e).

By this time, Edison's shareholders were growing increasingly nervous and agitated about their investments in the company. Though sales and peak load had continued their modest growth, the company's earnings had collapsed in 1990, 1991, and 1992 (see figs. 10.3 and 10.4). Earnings per share plummeted to $0.22 and $0.08 per share in 1990 and 1991 respectively. In the second quarter of 1990, Com Ed had to write off $208 million for Byron 1. Roughly two years later, it had to write off another $65 million. In the second quarter of 1990, the company recorded a reduction in net income of $253 million in response to the Supreme Court's December 1989 ruling. A year later, it wrote off $27 million for Byron 2 and Braidwood; it wrote off another $707 million in November of that year. The combined effect of these rate reductions and write-offs was to reduce the company's net income by $2.17 per share in 1990, $2.59 per share in 1991, and by at least $0.18 per share in 1992, with more likely to come.

At the time, I was an Edison shareholder, and I had become one partly in order to develop a shareholder's perspective on the company. I suspect that a shareholder (who had not been following Edison as closely as I have) would be growing deeply irritated at the management of our company. Reading excuse after excuse for the company's poor earnings over the past two to five years, such a shareholder would wonder why they allowed our company to get into this situation in the first place.

Despite these reductions in income, the company tried to maintain

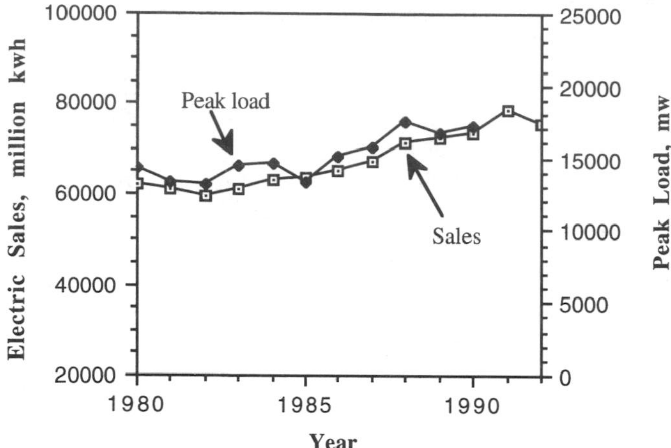

FIG. 10.3 **Edison's Sales and Peak Load Continue Modest Growth after 1982**

Sources: Moody's (1983, 1988, 1992).

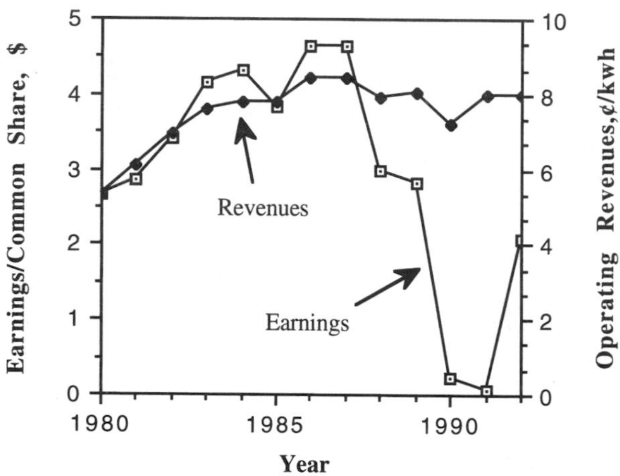

FIG. 10.4 **Edison's "Price" Declines Slightly after 1987, Earnings Collapse**

Sources: Moody's (1983, 1988, 1992).

its dividend payments of seventy-five cents per share. At first, it did so by drawing down on retained earnings. Given the magnitude of write-offs and impending rate reductions, however, that could not continue. Wall Street analysts began advising shareholders that a dividend cut was unavoidable and that it might be wise to sell. The market price of Edison's stock plummeted from $42 per share in December 1991 to just under $29 by mid-July 1992. At that point, James J. O'Connor told reporters that Edison would have to take drastic action to stop "a drastic deterioration in financial condition" (Maclean 1992a). A few days later, Edison announced that ICC and court decisions had "threatened the company's ability . . . to deliver reliable service" and that the company was going to reduce its workforce by 1,250 (from a total of 20,195), reduce its construction spending by $385 million, impose a wage freeze, and cancel all its television advertising. (Eight hundred of the 1,250 jobs to be eliminated were to come from the company's pool of 8,000 managers. Bide Thomas, the company's fifty-seven-year-old president, and three vice presidents soon chose to take the company's early retirement option.) O'Connor pinned the blame for the company's problems on a "breakdown of the regulatory compact . . . the long-held doctrine that investors will be allowed to earn a fair return on facilities authorized by the regulators to serve the needs of the customers" (Maclean 1992b). There is no such compact, Howard Learner retorted. The ICC has the duty to "allocate pain," he said, to distribute costs between the company, its shareholders, and its ratepayers. Furthermore, he and Susan Stewart of CUB were scarcely enthusiastic about Edison's cost-cutting moves. "As far as we are concerned," Stewart said, "the cuts are completely unjustified and constitute nothing more than economic blackmail to convince the Illinois Commerce Commission to raise electric rates in January" (Maclean 1992b). CUB's response deeply irritated Edison management. "We're not trying to scare anybody," O'Connor insisted, "we're not trying to bluff anybody." Claiming simply to be addressing a serious financial problem, he condemned Learner's and Stewart's comments as being "deceitful . . . wrong, misleading" (Maclean 1992c).

In mid-August 1992, the ICC staff recommended that Edison's rate be cut by 6.8 percent from the amount approved in 1990. In their view, Edison deserved an $87.2 million increase for its last three nuclear units, rather than the $482.9 million it had been authorized in 1991. The remaining $395.7 million would have to be rebated and refunded to Edison's customers. According to the staff, all of Byron 2 was needed, but only 19 percent of Braidwood 1 and 12.2 percent of Braidwood 2 were. Edison, which had just asked the Commission to raise the $483 million increase to $750.2 as planned in the 1991 order, wailed: the cut would result in "deeper cuts in our operations, with resulting impacts on customer service." Consumer groups, on the other hand, argued that Edison's rates should be cut by $506 million, or 8.7 percent, and that it should be ordered to refund $538 million to customers for savings that Edison obtained from a 1987 change in the federal tax law (Karwath 1992f).

Despite these growing financial pressures, the company tried to maintain its dividend. Finally, however, it could no longer do so. Just a few weeks after the ICC staff recommended a 6.8 percent rate cut in mid-August, Edison's Board of Directors decided to cut its dividend on common shares for the third quarter of 1992 from 75 cents to 40. Stock prices, which had fallen to just under $25 per share the day before, rose $1 in response. Learner and Stewart generally praised the cut, though they also suggested that a one-quarter cut would not be sufficient. Three months later, the Board of Directors decided to continue the 40 cents per share dividend.

In early October, the company made a potentially stunning announcement, one that might turn the modernist institution of regulated natural monopoly on its head in Illinois. Edison announced that it would need additional capacity by 1996 and that—rather than build that capacity itself—it was seeking proposals from independent companies to build and own a 400–500 megawatt nonnuclear plant, from which Edison would buy power. It had also asked factories, building owners, and other large customers to prepare conservation programs to help it save electricity during periods of peak demand. Those customers would receive lower rates in return. Unlike virtually every other Edison action over the preceding two decades, this action received *praise* from consumer groups. "To the extent that this represents a shift in strategic planning for Edison," Learner said, "as opposed to simply being a public relations strategy, it's good." But he also considered Edison's claim that it would need new capacity by 1996 to be "nonsense" (Karwath 1992g). Just a few weeks later, the company announced that it had decided to become an "energy service company" and to compete against other firms in providing a full range of energy services. "This is the most major change we've undertaken in the past 40 years," O'Connor proclaimed. "We don't have the option anymore to operate in a monopolistic environment. And we're not at all afraid of that" (Maclean 1992d).

Recall the brief discussion of "least-cost planning" in chapter 4. At the heart of least-cost planning is the idea that utilities can become energy service companies that sell electric power and energy conservation, trying to put the two on a "level playing field" and thereby minimize total societal costs of energy production and use. It is too early to tell whether Edison's commitment to becoming such a company is credible.

One week later, the company appointed a replacement for Bide Thomas. Announcing that "our industry is undergoing major change" and that "we want to be ahead of that change just as much as we possibly can," O'Connor introduced Samuel K. Skinner, outgoing chair-

FIG. 10.5 **Chairman O'Conner, Left, Introduces Edison's New President, Samuel K. Skinner**

man of the Republican National Committee, former White House Chief of Staff under President Bush, and former Secretary of the U.S. Department of Transportation (Maclean and Karwath 1992). O'Connor indicated that Skinner would be responsible for regulatory and legal issues, and Skinner said that his first priority would be to ensure that the company received the rate increases it had sought. Consumer groups expressed considerable skepticism about the appointment. A spokesman for Mayor Daley hoped that "maybe Mr. Skinner can help break the gridlock between Edison and its consumers," but Howard Learner noted that Edison's problems concerned rate issues, reliability, and nuclear power plant operation and that "none of those are Skinner's areas of expertise."

It is hard to tell whether Edison was trying to influence the commissioners by cutting its dividend, cutting its staff, hiring a new president, and proclaiming a change in corporate identity. Whether it was or not, the commissioners chose to decide in a manner counter to Edison's immediate self-interest. On January 6, 1993, six years after Edison and its political allies announced their "best deal for Illinois consumers," the ICC decided 5–1 to roll back the $483 million, 9 percent increase and to replace it with a new one of $152 million, or 3 percent.

In effect, therefore, the ICC cut Edison's rates by 6 percent. What is more, it ordered the company to refund to its customers the difference between the two rate increases. Totaling slightly in excess of $600 million, this refund was the largest in Illinois and possibly American history. The critical issue was, of course, the need for the last three nuclear units, and here the Commission disagreed only slightly with its staff. In the Commission's view, 93 percent of Byron 2 was needed, but only 21 percent of Braidwood 1 was. Braidwood 2 was not needed at all.

Stung by the Commission's order in what James J. O'Connor called "the rate case from hell" (Karwath 1993a), and claiming that the Commission had unfairly denied its common stock investors a fair return on their investments, Edison immediately appealed to virtually every court it could find, seeking a temporary stay in the Commission's order. In their view, the ICC had adopted "poor public policy" that confiscated Edison shareholders' investment without due process. "They need some help over there frankly," said Samuel Skinner, Edison's new president, in disgust. "That is outrageous," cried ICC Chairwoman Craig. "That is what they have done in the past, and that is where they [Edison] and the commission have gotten into trouble" (Maclean and Karwath 1993). Nothing, she said, prohibits the company from applying for a new rate increase. Furthermore, the most recent data about electric power demand (rather than the data used in this particular case) would probably show that all or most of the last two units are used and useful. Despite Craig's admonitions, Edison appealed to the Illinois Supreme Court for an emergency stay of the order, but the Court issued a one-sentence order refusing to delay the cut and refund. It then turned to the U.S. Supreme Court to delay the cut and refund on the grounds that the ICC had violated the company's right to due process, but Supreme Court Justice John Paul Stevens denied the request. Howard Learner of BPI and Susan Stewart of CUB said that Edison was taking "a desperate frolic and detour" and was "really grasping at straws" in appealing to the U.S. Supreme Court, and that "Edison should stop wasting its time blaming other people for actions that are its own responsibility" (Karwath 1993b). O'Connor warned that the company would have to make additional cost cuts if a larger increase was not granted. Editors at the *Chicago Tribune* strongly supported Edison in its appeals, claiming that "the increasingly bitter and politicized debate sometimes has obscured the fundamental need to keep Edison a financially viable and dependable electricity supplier. . . . [C]onsumers should worry about who will pay Edison's higher financing costs and what they'll do if their air-conditioners don't work on some sweltering July day" (January 10, 1993).

Finally, on January 21, 1993, Edison succeeded in persuading the Illinois Supreme Court to delay implementation of the rate cut and refund pending results of Edison's appeal. The "rate case from hell" continued its demonic journey through the crippled modernist institution of regulated natural monopoly.

Interlude

In a Classroom at the University of Iowa

Back in Iowa City, teaching on another beautiful spring day. None of us want to be in the classroom on a day like this. Still we all want to be here, whether out of a lust to learn, a desire to change the world, or the hope of finding a good planning job. The students and I have been reading and discussing draft chapters of this book as part of the course, and now we're nearing the conclusion. What to write in the final chapter? Though the temptation is strong, I do not want to stand metaphorically on the "central plateau" acting as if I possess knowledge and that the students and other readers are merely empty vessels, waiting to be filled. Prior experiences have taught me that I learn from them every day. Garth challenges me to defend my use of "modernist" and "postmodernist," arguing that those labels are unnecessary and counterproductive; Mike urges me to think of Edison as a corporation that is trying (perhaps none too successfully) to market its products and plans; Julie jokes that she "wants to say something smart but can't," then argues that the radical difference in tone between the interludes and the chapters might alienate some readers; Sean worries that the alphabet soup of acronyms, technical jargon, and characters might confound the readers' ability to understand what is going on; all of them feel that the book really starts coming to life and sustaining its claims about persuasive storytelling in chapter 5. Worn by work and complicated personal problems, but feeling recharged by the students and wanting to finish this book I started over six years ago, I return to my metallic gray desk to ponder the final chapter. The cursor blinks, awaiting my words.

11

The Plateau in the Web: Planning as Persuasive Storytelling within a Web of Relationships

Eighty years ago Daniel Burnham told the proto-planners of his time to "make no little plans." "Aim high in hope and work," he said, "remembering that a noble, logical diagram, once recorded, will never die, but long after we are gone will be a living thing, asserting itself with ever growing insistency" (Wrigley 1983, 71). Six decades later, just as urban and regional planners were turning away from the modernist pretenses of rational planning, and shortly before the OPEC oil embargo radically altered the nation's economic course, Commonwealth Edison embarked on a nuclear power expansion plan that was by no means little. Trusting in the power of Reason, and wanting to ignore the particularities of local politics, Edison's electric power planners continued the modernist effort to impose Man's (or, in this case, Edison's) will on the future. Edison's troubled experience with that plan marks a decisive turn away from modernist planning and towards a form of planning for which we do not yet have a name. Though it lacks a name, and probably does not need one, this form of planning is rooted in persuasive storytelling and the rhetorical construction of communities.

Edison's Costly Descent from the Central Plateau of Modernist Planning

Let us briefly recall how Commonwealth Edison's managers and planners sought to reenact the modernist story of planning. Comfortably secluded in their downtown Chicago headquarters, surrounded by the electrically illuminated and air-conditioned towers of Burnham's successors, Edison's professional managers configured their electric power system as a vast and immensely complicated machine. Contemplating the gleaming towers, they defined themselves collectively as a rational decision maker who could make the machine progressively more efficient and then impose the requisite changes on the machine's environment. Gazing down upon the deteriorating neighborhoods that spilled from the gleaming towers, down where the company's machine wove its way through the physical and social landscape, they could just barely see all the little people scurrying through the sights and smells and sounds of the dark wilderness of Chicago politics. Though largely emotional and irrational, these little people could gradually be trained to interact harmoniously as unthinking components of the machine. To become, if you will, the "innocent little girl" that appeared in one of the company's advertisements (see fig. 10.2).

Up there, in the towers' rarified air, it was possible to take the modernist story seriously. Let the data speak. Construct the models. Forecast future demand for electric power with the best methodologies and analysts that money can buy. Use those forecasts to determine how much new generating capacity is needed. Identify the least costly way to generate that new power. Write a plan for building those plants. Provide the Commerce Commission with the evidence that justifies building new plants and reducing consumer rates. Build the plants. Observe the price of power decline and the demand for it respond by increasing. Start the cycle anew, pleased that shareholders and consumers will both benefit. Recognize the truth of architect George B. Ford's words back in the twenties: if the task of planning is approached scientifically, then "one soon discovers that in almost every case there is one and only one, logical and convincing solution of the problems involved" (Foglesong 1986, 213).

The data said that demand for electric power in Edison's service area had been growing at an average annual rate of 7.8 percent from 1961 through 1973. The data said that peak load—which stood at 10,943 megawatts in 1970—would grow to 17,850 megawatts in 1977 and that the company would need to add almost 7,000 megawatts of additional generating capacity to meet that demand. The data showed that nuclear power would provide the least costly way of generating that additional power. Consequently, the one best solution to the problem (or opportunity) of rapidly increasing demand was for Edison to build six new nuclear power generating units, each with a capacity of 1,100 megawatts, at its LaSalle, Byron, and Braidwood stations. Cost estimates showed that construction of those units would be completed by 1980 at a cost of $2.506 billion. Net present value revenue projections showed that the rates (prices) consumers pay for electric power would decline as the plants began generating useful power.

Confident in its forecasts and plan, Edison began building its six nuclear units in the early 1970s. However, the company soon encountered a series of difficulties (all due, of course, to factors beyond its control) which endangered that plan. Rather than growing at 7.8 percent, peak demand grew at an average annual rate of 2.3 percent from 1974 through 1988. Rather than growing by 6,900 megawatts between 1970 and 1977, peak load grew by 2,990 megawatts. (As of 1990 it still had not reached the load Edison projected for 1977.) Rather than costing approximately $400 per kilowatt, Edison's last six nuclear units turned out to cost almost $2,100. Rather than declining, the average rates that Edison charged its customers (measured in terms of gross operating revenues divided by kilowatt-hour sales) increased from 4.51¢/kWh to 7.35¢/kWh in constant 1982 dollars. Despite these difficulties, the company did finish building its six nuclear units. But it did not finish them until 1988 at a total cost of $13.7 billion, and along the way it had to increase its rates by 63 percent (in constant 1982 dollars), saw its earnings drop from $4.75 per common share in 1987 to $0.08 in 1991, and had to cut its dividends from $3 per common share in 1987 to $1.60 per share in 1991.

Rigidly focused on completing the company's troubled expansion plan, on cutting a clear path through the wilderness below, Edison's managers found that they could no longer simply contemplate and control the future from their central plateau. To their dismay, the plan and its tidal wave of rate increases motivated consumer, environmental, antinuclear, community groups, and sympathetic elected officials to coalesce in opposition (see fig. 11.1). Instead of simply presenting evidence that justified the plan, Edison's planners and managers had to descend from their plateau and then try to persuade specific audiences at specific times in specific contexts, in the face of determined opposition.

Commonwealth Edison's descent from the central plateau of modernist planning forms the heart of this book. In order to complete its expansion plan, Edison had to explain why the plants were needed, why their construction was being delayed, why their cost was increasing so dramatically, why rate increases were justified, and why the City of Chicago should not leave Edison's system. In other words, the company had to tell a continually evolving story about the past and about the future. This effort at persuasive storytelling proceeded along two parallel paths. One track led to the familiar institution of regulated natural monopoly. There Edison relied on the normal discourse of

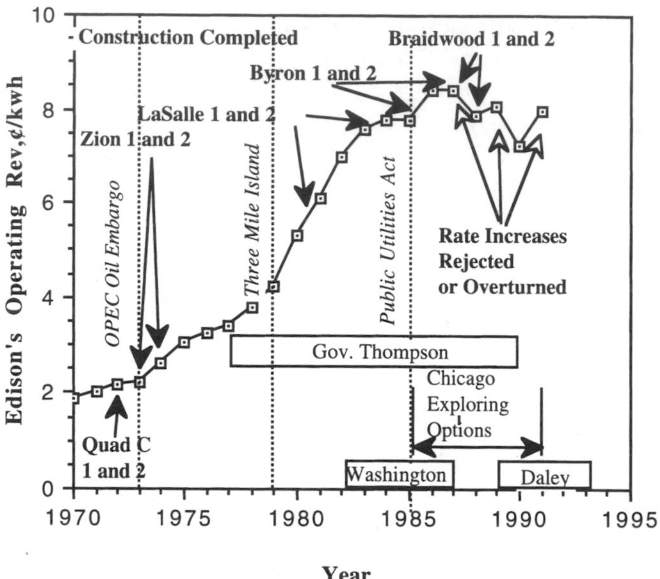

FIG. 11.1 **Edison's Tidal Wave of Rate Increases Meets Stiff Resistance**

public utility economics, seeking to persuade the ICC commissioners and the courts to embrace Edison's story and reject CUB and BPI's alternative. The second led (from Edison's point of view) into the wild chaos of Chicago politics. There they had to speak in the abnormal discourse of threats, promises, public utility economics, and advertising and thereby convince the mayors, aldermen, and people of that city not to abandon the company's system. In both cases, Edison's managers remained inspired by modernist pretenses and asserted their expansion plan with ever growing insistency.

At the core of Edison's effort to persuade the ICC was the company's claim that the delays, cost overruns, and rate increases were due to *factors beyond its control.* At the heart of its effort to persuade Chicagoans was its claim that precinct captains (corrupt and incompetent politicians) should not be trusted to manage a highly complex—*and potentially quite dangerous*—nuclear-based electric power system. These two claims are closely interconnected and deserve our careful consideration.

Modernist planning presumes first that planners can disaggregate the rich complexity of life into discrete and quantifiable variables and then relate relevant variables to one another in a model that coherently and truthfully represents reality. It then presumes that the planner (or those for whom the planner works) can control (or at least confidently

TABLE 11.1 **Five Rate Increases Denied, Overturned, or Pending Appeal from 1985 through 1993**[a]

Date of ICC Decision	Rate Increase Sought/Approved ($ million)	Rate Increase Sought/Approved (%)
Oct 24, 1985	544.6/494.8[b]	12.2/11.0[b]
Jul 16, 1987	660.0/_____[c]	13.3/____[c]
Dec 30, 1988	1,414.9/480.0[d]	26.9/8.0[d]
Mar 8, 1991	1,231/750.0[e]	23.0/14.0[e]
Jan 6, 1993	982/152.0[f]	17.7/3.0[f]

[a] *Sources:* ICC 1985, 1987, 1988, 1991, 1993.

[b] The increase was phased in over four years. Rates would increase by 9.2% in 1985, rise to 12.0% in 1987, then decline to 11.0% in 1989. The Illinois Supreme Court subsequently overturned the decision on June 16, 1987.

[c] The Commission rejected the rate increase associated with this "negotiated settlement."

[d] The Illinois Supreme Court overturned this "settlement agreement" on December 21, 1989.

[e] The ICC decided to phase the increase in over four years. Rates would increase by 9.0% in 1991, rise to 13.3% in 1992 and to 17.8% in 1993, then decline to 14.0% in 1994. The Illinois Supreme Court overturned this decision on December 16, 1991.

[f] The net effect of this decision was to reduce the increase approved in 1991 by 6 percent in the first year. Edison appealed the decision, and the Supreme Court allowed the first phase of the 1991 increase to remain in effect until the appeal was heard.

account for) those variables. Contrary to those presumptions, Edison retroactively claimed that several variables were beyond its control and behaving unpredictably, including the 1973 OPEC oil embargo, price inflation, inflation in the cost of capital, the 1979 nuclear accident at Three Mile Island and subsequent NRC regulations, the 1986 nuclear accident at Chernobyl, and—most important for the present book—the ability of consumer and other groups to affect electric power legislation and regulation. Surely some of these variables were beyond the company's control, though that by itself casts doubt on modernist planning. But I want to draw attention to the last factor. Consumers, antinuclear groups, environmentalists, and supportive elected officials do not appear in Edison's models and forecasts. By not including those groups in its models, the company implicitly assumed that they were (or should be) irrelevant agents. But in practice they proved far from irrelevant. To Edison's dismay, consumer groups influenced Illinois legislators to enact the 1985 Illinois Public Utilities Act and then persuaded the ICC or the courts to reject or overturn a series of proposed rate increases (see table 11.1).

The company obtained a $495 million rate increase for its Byron 1 plant in 1985, but consumer groups and governmental agencies persuaded the Supreme Court to overturn that increase a year and a half later. The company proposed an innovative $660 million increase in late 1986 as part of a "negotiated settlement" that provided the "best deal" for consumers. Seven months later consumer groups persuaded the ICC to reject that proposal. The company applied for a $1,414 million increase in mid-1987. The ICC approved a two-step $480 million increase in late 1988, but consumer groups persuaded the Supreme Court to overturn that increase in late 1989. The company applied for a $1,231 million increase in 1989. The ICC approved a multi-phase $750 million increase in March 1990, only to have the Supreme Court reject that decision in late 1991. Upon reconsideration, the ICC approved a $152 million increase in January 1993. The company considered the increase to be punitively small and appealed the decision to the Supreme Court. As a result the utility had—by mid-1993—still not been able to obtain a legally defensible and permanent rate increase for any of its last three units. The tidal wave of rate increases that produced a cumulative increase of 63 percent in constant dollars between 1970 and 1988 had been blocked, or at least temporarily impeded by political actors who did not appear in Edison's modernist planning models.

To maintain control over their electric power machine, Edison's managers also had to persuade the mayors and aldermen of Chicago to renew the company's franchise with the City. Strutting like soldiers in rigidly organized battalions, turning in unison to authoritatively sounded commands, Edison's managers, planners, and negotiators marched into the dark wilderness of Chicago politics. Endangered and perhaps frightened, they engaged in persuasive tactics (advertising, threats, promises) that their opponents considered duplicitous, manipulative, and untrustworthy and that ran directly counter to the company's modernist assertions. Claiming "we're there when you need us" and that the company was against people who want to "get rid of us," its leaders gave the marching orders and barked the commands: "We own the highway," one shouted. The City is "being led down the garden path," cried another. Falling in line, others cried: "a few bureaucrats and activists" are trying to "take over . . . your electric system," "precinct captains would be running the power company," and "STOP THE BUREAUCRATS AND ACTIVISTS NOW!"

At first the battle lines seemed clear. Led by the charismatic Mayor Washington, a progressive coalition of community and consumer groups firmly stood its ground. Stunned by Edison's recent and prospective rate increases, and by Edison's obstinate insistence on building unneeded and dangerous nuclear power plants, Washington's coalition feared that the rate increases would "deliver a crippling blow to the economic health of Chicago." Firmly backed, the Mayor declared Edison's 1987 best deal proposal to be "the most outrageous railroading job ever perpetrated," said the later proposal to raise rates by 27 percent "represents a tidal wave that threatens to wash the city's economic base to the sea," and considered "dropping off Edison's system." But then the charismatic Mayor died, his coalition fragmented into many small and poorly coordinated bands. Though attacked from many sides by these guerrilla-like groups, Edison's battalions marched straight ahead ("we're trying to show that we have some leverage too," said Edison's chief negotiator) and pummeled its opponents into bitter submission. In the end, the company could not persuade Mayor Daley ("I'm no wimp") and the City Council not to terminate the franchise in late 1989, but it was able to persuade the Mayor and Council to agree to a new franchise in late 1991 that extended Edison's ability to "serve" the City for the next twenty-nine years.

Edison claimed that city politicians and bureaucrats could not and should not be trusted to operate nuclear power plants. It also claimed that the construction delays and cost increases were caused by factors beyond the company's control. What connects these two claims? Simply this: the narrative of modernist planning. At heart, the company's managers believed that they and other technical experts could design, build, and safely operate nuclear power plants at a cost measurably below other viable generating alternatives but that politically inspired meddlers had overreacted to the 1979 accident at Three Mile Island and had successfully pressured Congress and the Nuclear Regulatory Commission to impose a new set of regulations which dramatically and unnecessarily inflated the cost of Edison's last six nuclear units. Those same meddlers had conspired to persuade the Illinois General Assembly to enact a new Public Utilities Act in 1985, an act that was loaded with features hostile to utilities. The dark wilderness of Chicago politics extended, in Edison's view, to the labyrinths of Congress, the General Assembly, and the bureaucracies of Washington and Springfield. Note, however, that Edison could not simply assert its modernist story about scientific expertise, nuclear power, and political meddling. It had to persuade Congress, the NRC, the General Assembly, the ICC, and the mayors and aldermen of Chicago that the company's story made more sense than alternative stories. Most important, Edison had to alter its original story to account for the Three Mile Island and Chernobyl accidents, the waste of more than $1.7 billion in building the Byron 2 and Braidwood units, and the surprising series of blackouts that struck Chicago in 1990.

The deeper Edison's managers plunged into the dense thicket of

Chicago politics, the more they had to toss off the cloaks of modernism while claiming to embrace them. In the end Edison managed to maintain its status as a franchise-limited regulated natural monopoly. But it did so only by reenacting the political origins of the modernist planning story and by weakening itself in the process.

Planning as Persuasive Storytelling within a Web of Relationships

If modernist planning (along with its pretensions about objectively forecasting and controlling the future) is no longer viable, what can take its place? Is planning itself so intimately tied to modernist thought that it too is necessarily bankrupt, or can we reconfigure planning in a manner that is more consistent with the temper of the times? Throughout this book, I have suggested that planning can best be understood as a form of persuasive and constitutive storytelling. I now want to connect this suggestion with the notion (first introduced at the end of chapter 1 and the start of chapter 2) that cities can best be thought of as nodes in a global-scale web of links and flows.

Edison's modernist planning story set the company and its electric power machine apart from its environment (from all that surrounded it). Imagining that the city below could be treated as a featureless plain upon which the company's plans could readily be imposed, Edison's managers rejected an alternative configuration: that the central plateau is itself part of the landscape of planning, that the plain is not featureless but a densely patterned web of relationships whose strands extended far from Chicago, and that embeddedness in the web implies a radically different approach to planning. By claiming to be apart, Edison could only talk in terms of factors within or beyond its control, but by blaming factors beyond the company's control for construction delays and cost overruns, Edison exposed its embeddedness in the web. Firmly attached to the modernist conception of planning, it could only respond to that web by claiming to plan "scientifically" but all the while acting "politically."

So let us reject the modernist imagery of plateau and plain, and reconstruct the story about Edison's nuclear expansion plan in light of the web metaphor (Landow, 1992). In this view, the Commonwealth Edison Company, the City of Chicago, the Illinois Commerce Commission, and the other organizations discussed in this book can be thought of as *nodes in a global-scale web,* a web that consists of a highly fluid and constantly (albeit subtly)

Irony again. Was it not large-scale technical systems (telephone and television, natural gas and petroleum, airline, computer and electric power) that constructed these

webs in the first place? Much as centralized electric power systems helped to disperse and socially fragment society, so too rational electric power planning has helped create the conditions for a postmodernist form of planning.

changing set of relationships. These relationships in turn can be defined as links between nodes, as paths through which goods, services, energy, capital, information, and other social exchanges flow. Edison's electric power system—and the franchise agreement that established Edison's and Chicago's rights and duties—can be understood as one of the links in that web. In fact, the schematic map in fig. 3.1 conveys that image quite directly, as does the comment by Edison's Vice President Donald Petkus that "we own the highway." The ICC's regulatory powers over Edison can be thought of as another link in the web. In this instance the potential fluidity of relationships becomes quite clear. Recall the "regulatory compact" that used to define the rights and duties of a regulated natural monopoly (see chapter 3), and note Howard Learner's statement in chapter 10 that there no longer is any such compact.

This global-scale web, so intricate and extensive and fluid, can be likened to a text that can be constructed and read in multiple ways. Plans, in turn, can be understood as persuasive stories about how particular nodes or links in the web should or will change in the future. Edison constructed one interpretation of the web, an interpretation based on the company's proclaimed ability to contemplate and control from the central plateau. Based on that interpretation, the company then authored a nuclear expansion plan, a "negotiated settlement," and other planning documents that specified how much better things would be if the company's plans were adopted and pursued. It—and its opponents—continued to use standard planning tools: models, forecasts, surveys, and the like. But note the fundamental shift that occurs when one stops thinking in terms of the plateau and the plain and begins thinking in terms of the web. Given the web, the issue becomes not so much the absolute Truth or Goodness of forecasts and other planning statements, but the persuasiveness of their construction and interpretation and the openness of each node to alternative rhetorics.

In the particular nodes currently named Commonwealth Edison and the Illinois Commerce Commission (which together I have referred to as the institution of regulated natural monopoly), people are permitted to speak only the normal disciplined discourse of public utility law and economics. They must talk in the strange language that we first encountered in the prelude to this book: net present value revenue requirements, prudently incurred and reasonable costs, reasonable rate of return on common equity, AFUDC, CWIP, used and useful plants, and so on. Convinced that their way of talking leads to Truth and Goodness, Edison's managers and the ICC's commissioners and staff radically translated and edited contrary rhetorics—such as those spoken by consumer and community groups—before allowing them in. Those rhetorics were, after all, merely the incoherent babblings of self-interested meddlers or the undisciplined ramblings of people who do not have the expertise necessary to manage and regulate such a complex enterprise.

Note, however, that such talk, which appears to be incoherent or undisciplined from the point of view of Edison and the ICC, did not simply disappear. The CEOC and other community groups live and work in a city that is evidently decaying—despite Edison's efforts to characterize Chicago as a world-class metropolis whose growth depends on the company's electricity—and those groups speak a rhetoric which expresses their sense of anger and bitterness at that decay. Edited out in the institution of regulated natural monopoly, the community groups' rhetoric of moral indignation reappeared in the struggle over Chicago's effort to explore alternatives to Edison. And, unlike the nodes of Edison and the ICC, Chicago was far more open to diverse rhetorics and stories. Some characters continued to speak the rhetoric of public utility law and economics, but others spoke the moral language of right and wrong and the political language of favors and deals. Rather than being edited at the border of the node, those rhetorics and stories (which reflect the ways people speak in other nodes of the web) interplayed in rich and often confusing complexity.

Each character in my narrative about Edison's expansion plan and Chicago's exploration of alternatives tried to tell a persuasive story. In each case, they drew attention to linkages and nodal boundaries that supported their versions while suppressing ones that did not. Edison constructed a modernist tale: they looked down on the plain, saw irrational little people being pushed this way and that by political bandits, and they ordered their battalions to clear a path through the wilderness in the name of order and rationality. According to this tale, the boundaries and links of the web were clearly definable; the network could be conceptualized in terms of discrete, quantifiable, and independent variables. Yet Edison's planners and managers seemed to know that somehow that was not the case (*factors beyond our control . . .*). Glancing upward at Edison's tower, the guerilla bands in the wilderness also constructed a modernist tale, only this time with a different plot and different set of characterizations. They saw an arrogant and domineering company that sought to impose its will on others; a company that claimed to act in the interests of consumers when consumers themselves were, for the most part, deeply upset at the company's actions; a company that condemned activists and bureaucrats for acting politically, all the while acting politically itself.

Recalling the previous chapters, it seems clear that the web has no objective structure or meaning. Persian Gulf oil exporters, nuclear plant technicians in Pennsylvania and Ukraine, bureaucrats in Washington, consumer group lawyers in Chicago, judges in Springfield, shareholders spread across the nation, and many others all had a hand in determining the costs and timing of Edison's nuclear plant construction program. Given that web of interconnected relationships, it no longer makes sense to assume hard-and-fast nodal boundaries or linkages. What does make sense, and what we have seen happen repeatedly in the present story about Edison's expansion planning, is that the company, its opponents, and all other participants (myself included) have constructed boundaries and then attempted to narrate persuasive stories in terms of those initial constructions.

Each effort to narrate a persuasive story about the nodes and links of a web is necessarily complex, incomplete, fragmented, and conflictual, but the effort to tell a story (and to read critically the stories written by others) necessarily reveals more and more about the web: about the diversity of speakers/audiences, about the diversity of interpretive communities and rhetorics, and about the diversity of meanings. But, firmly attached to their own narratives, none of the characters in the Edison/Chicago story seemed capable of responding to this fragmentation and diversity. Potentially fluid, the web seemed ossified by language, by the stories Edison and others told about it.

When one plans, one plans as part of a web of relationships. However, the point of view from which one plans varies with one's location in the web. Viewed from another part of the plain, from another node in the web, from a university in Iowa City, Edison and its opponents appeared to be locked in a frozen embrace, sharing conceptions of the plateau and the plain, of modernist planning and politics, while disagreeing about plots and characterizations.[72] Edison was engaged in a heroic effort to bring truth and order into a flawed world; the consumer groups and their allies were engaged in a romantic quest to overcome evil. Good versus bad, truth versus falsehood, regardless of point of view. Observing this frozen embrace from another part of the web, I see a story that has turned tragic. Edison and its opponents are devouring one another and—by doing so—opening the door for some other node in the web to supplant them both. I am reminded of two historical analogies. Fourteen centuries ago, the Byzantine and Parthian Empires fought one another to exhaustion on the Mesopotamian plains. Exhausted, neither empire was able to withstand the totally unexpected rise of the Mohammedan Empire. Similarly, the United States and the Soviet Union fought a cold war for over three decades. In the end, many now say, Japan won.

The key to melting the frozen embrace, to unlocking the grid, is therefore to appreciate the rhetorical nature of planning in contemporary America. To plan effectively, we have to learn how to take rhetoric seriously. We have to learn how to construct persuasive stories about the future. We have to recognize that we are embedded in an intricate web of relationships, that we have to construct understandings of that web, and that we then have to persuade others to accept our constructions. This does not mean abandoning modeling, forecasting, surveying, and the other quantitative tools of planning; it means embracing the realization that other groups and organizations will construct their models differently, will use different assumptions when forecasting, and will ask different questions when surveying. A rhetorical approach to planning does not mean abandoning claims that certain actions are morally wrong, either in themselves or in their consequences; it does mean embracing the realization that other people will make strong and often completely contradictory moral claims. Nor does it mean abandoning the claim that a certain course of action might be more effective politically than some other; it just means that opponents will argue differently.

Given such a rhetorical conception of planning, we should not be surprised or offended by the discovery that Edison and its opponents planned (and acted) technically and politically. Nor should we be surprised that they constructed alternate stories about Edison's expansion plan, about its motivations and likely consequences. What we should do—rather than feign surprise—is find ways to set these alternate stories side by side, let them interact with one another, and thereby let them influence our sense of how particular nodes and links in the web should change, are likely to change, and why.[73]

I have tried to exemplify just such a rhetorical conception of planning in this book. Writing as a peripheral first person narrator and using interludes and inserts, I too have come down from the "central plateau" and become a part of the story. Rather than attempting to narrate an objective history of Edison's nuclear expansion plan, as if I somehow had unique access to all the nuances and intricacies of the story, I have sought simply to narrate a persuasive tale that truthfully represents my understanding of planning in general and Edison's expansion planning in particular, and which, I hope, demonstrates the importance of quality reading and quality listening in planning. But make no mistake. I do not claim to have told the only possible story, or to have been neutral and unbiased in constructing my version of the tale. My tale of gridlock is constructed from the point of view of a peripheral first-person narrator who currently lives and works 240 miles west of Chicago but who also is conscious of having established several specific relationships with other nodes in the web. As a scholar at the University of Iowa, I have read widely in the fields of urban and regional planning and electric power policy making. I have also presented several scholarly monographs at national and international conferences and then exposed my work to critical peer review in refereed publications. As a resident of Hinsdale, Illinois, from 1983 through 1986, I have been one of Commonwealth Edison's customers and a member of the Citizens Utility Board. As the author of this book, I became an Edison shareholder and a member of the Chicago Electric Options Campaign, and I have spent hours directly observing Chicago planners, aldermen, community groups, commissioners, and lawyers. As the real flesh-and-blood person revealed in the interludes, my sympathies clearly lie with the community and consumer groups. Even so I have tried to be fair to Edison, and I see limits to truth and goodness on both sides.

To act wisely and effectively in a context of postmodern moral and factual ambiguity requires a considerable degree of moral courage. Somehow planners and advocates must learn how to persuade yet be open to persuasion, and they must strive to create forums in which that kind of mutual persuasion can occur. But how can such forums be created, and what would be their defining characteristics?

At the end of chapter 9, I presented a hypothetical story that planners and advocates in the progressive coalition might have authored in an effort to persuade Chicagoans to turn away from the status quo of regulated natural monopoly. That story presumed that those planners and advocates were (and should be) deeply committed to the values and norms of their community and that they sought to construct a world that was more in accord with those commitments. But that hypothetical story did not permit the members of the progressive coalition to remain inflexibly tied to those values and norms. Rather, it constructed a future in which the coalition members engaged their opponents (who were, after all, the coauthors of the story) in dialogue. They listened to Edison's modernist narrative of planning and control, took it seriously, and let themselves be changed through Edison's reasonable expectations and arguments. Edison's claim to do whatever they wanted because "we own the highway" was not reasonable; its complaint that the progressives "want to get rid of us" was.

So the planners and advocates of the progressive coalition would have been (at least in my thought experiment) open to transformation through the encounter with Edison. Such openness would have required considerable moral courage, for any members of that coalition who made such a move would surely have been accused of being naive, utopian, foolish, or worse yet, back-stabbing traitors. So it is not a move to be taken lightly. I should think such a move would be taken only with forethought and consultation, ensuring that friends were fully aware of what was being done and why. Moreover, there would have to be some reasonable expectation that such a move might be met with reciprocal moves on the part of Edison's managers.

My hypothetical story also presumed that it would be possible to conduct an open, public, and democratic discourse about Chicago's electric future. In my thought experiment, that discourse would have occurred within and before the Mayor's Task Force on Energy and—more importantly—in community forums held throughout the city. What would such forums require, in order not to degenerate into shouting matches between competing interests, back-slapping pep rallies, or rooms full of empty chairs and blank stares?

To answer this question, I want to return to Leith and Myerson's (1989) advice that we think of persuasive efforts as involving a flow of utterances, replies, and counterreplies, wherein the meanings of those utterances are open to diverse interpretations. Given that advice, public and democratic forums would require (1) conversation in ordinary English (rather than the "normal discourse" of particular interpretive communities), (2) advance distribution of written plans and analyses, (3) oral presentations of those plans and analyses to the public, such that members of the audience would have an unencumbered opportunity to ask questions and to present alternative interpretations, and (4) written responses to those questions and alternative interpretations. But these recommendations do not go far enough, for the present book amply documents the extent to which wealthy corporations can shape the flow of argument and the public's attention and understanding of particular issues. Wealthy corporations simply find it too easy to purchase opportunities to tell their story over and over again in the mainstream media. To truly engage in a public and democratic discourse relevant to nodes in a global-scale web, planners and advocates and others will have to turn away from the mainstream news media and begin designing, promoting, and using advanced interactive media that permit authors and readers, planners and audiences to exchange ideas and information, values and preferences, hopes and dreams. In a word, to tell their own stories, rather than have those stories be drastically edited, clumsily translated, and awkwardly narrated by someone else.

The challenge for contemporary planners in northern Illinois and elsewhere is, therefore, not so much to explore alternate paths into the future, or even to invent alternate paths. The challenge is to drop the imagery of the plateau and the plain; to begin planning based on the imagery of webs, nodes, and links; to find ways to construct stories that reconfigure the web in persuasive and compelling ways; and to construct new forums which enable public and democratic argumentation. Thus at the end of this extended tale I find myself revising Daniel Burnham's advice to modernist planners: Spin no sterile yarns, my fellow planners, for they do not provide us with the passion, reason, and power needed to construct our paths into the future.

Postlude

E-mail to a Friend

Date: Fri, 8 Jul 1994 10:08:02
From: Jim Throgmorton ⟨jthrogmo@blue.weeg.uiowa.edu⟩
To: colleagues and friends
Subject: end of the book

Great news! I've finally finished the book. It's amazing how long one of these can take, given all the reviews, editing, and the like. I know that some of you have already read or heard parts of it, but I'm really curious to know what you think of the whole thing. Does it contribute to the larger public discourses about planning and expertise, about the risks and benefits of nuclear power, and about the politics of determining what's possible and desirable in specific localities at specific times? Does it help us understand the possibilities of planning in contemporary America? Please let me know what you think.

Oh yes, I almost forgot. You are surely wondering what happened to Edison's persistent effort to obtain a legally defensible rate increase for its last three nuclear units. Rebuffed in their appeal of the ICC's January 1993 rate order, and feeling paralyzed by the cloud of uncertainty and delay, Edison's managers decided to negotiate in good faith with consumer groups and governmental agencies. Late in September of that year, the contending parties agreed upon a settlement: Edison would refund $1.34 billion to its customers and reduce its average rates by 6 percent; in return, the consumer groups and other intervenors would drop the many cases that were then active in the courts or before the ICC. The parties also agreed that consumers would receive their entire rebate before any new rate *increase could take effect. Though quite pleased overall, consumer groups avoided committing themselves to any future rate increase. "No secret deals. No 'nod nod, wink wink,'" said Howard Learner. "We're not going to lie down if there's a new rate filing by Edison" (Karwath 1993c). Five months later, after fulfilling the settlement's provisions, Edison surprised no one by applying for a new $460.3 million (or 7.9 percent) rate increase. The Commission surprised few when it approved a $303 million increase in January 1995. CUB appealed the decision. Here's a photo from the September 1993 announcement. I thought you might enjoy it.*

FIG. 11.2 **Chairman O'Connor (left), Howard Learner, and Susan Stewart Announce Rate Settlement in September 1993**

Notes

Chapter One

1. The roots of this core idea (which I refer to as modernist planning) can be traced back to the Enlightenment of the seventeenth and eighteenth centuries and the subsequent rise of technocratic consciousness. Enlightened by "subject-centered reason," humans would be freed from ignorance, superstition, and the authoritarian requirements of tradition. Committed to thinking about nature and society in a new and more scientific way, they could use more "rational" methods to deal with the problems of human society. Nation-states could then use these more "rational" methods to guide the future development of their cities and economies. As Toulmin (1990, 75) puts it, the modernist view claimed that "the validity and truth of 'rational' arguments is independent of *who* presents them, *to whom,* or *in what context.*" Thus conceived, modernist planning presumed rigid boundaries between fact and value, rationality and emotion, humans and nature, and experts and laypeople. Though I focus on planners and policy analysts in this book, it is important to understand that the points I make apply more generally to any group of professional experts who rely on the rhetoric of science and who believe that public policy should be guided by expert advice. See Friedmann 1987, Habermas 1987, Hawkesworth 1988, Fischer 1990, and Toulmin 1990.

2. For a complementary history about the rise of technical rationality in American politics, see Fischer 1990.

3. To the best of my knowledge, that history has not yet been written. However, there have been many fine historical studies about various aspects of the industry, most notably McDonald 1957, Hubbard 1961, Funigello 1973, DuBoff 1979, Hughes 1983, Rudolph and Ridley 1986, Nye 1990, and Platt 1991.

4. The telling of the past in interaction with a community of readers presumes that the narrator writes from the point of view of a definable tradition. As the hermeneutical philosopher Hans-Georg Gadamer puts it, "we are al-

ways situated within traditions. . . . tradition . . . is always a part of us, a model or exemplar" (1991, 282). Furthermore, "[t]o enter into a practice is to enter into a relationship not only with its contemporary practitioners, but also with those who have preceded us in the practice" (MacIntyre 1981, 194).

5. Philip Langdon (1994, 77), for example, writes that "planners lost much of whatever competence they once had as makers of physical form. There are still many individuals called planners, but most of them might more accurately be termed application-accepters and permit-dispensers."

6. This systematic lack of attention also derives from the fact that most electric power is produced at generating facilities located well outside large urban areas, with the primary fuels used to generate that power being extracted from coal mines, uranium mines, oil and gas fields, and hydroelectric dams located even farther away. The end result is that the social and environmental costs and consequences of electricity use are largely invisible at the point of consumption (Stern and Aronson 1984).

7. Prior to 1895, factories had to be located near primary sources of power. In that year, however, the utility serving the Buffalo, New York, area first transmitted electric power over a long distance (13 miles) via alternating current. From that point on, new factories were free to locate in much more dispersed patterns. Similarly, from 1909 though 1919 electric motors increasingly displaced water and steam as the primary forms of generating industrial light and motive power. This new technology facilitated mass assembly lines and induced new factories to locate in open suburban areas. (See DuBoff 1979 and Nye 1990.)

8. Note the temporal correlation between this academicizing of planning discourse and the eradication of older city neighborhoods through urban renewal. According to Jacoby (1987), "public intellectuals" (writers and thinkers who do not simply cherish thinking and ideas, but seek to contribute to open discussions with a general and educated audience) thrived in the urban bohemias that were systematically destroyed in the name of urban renewal and interstate highways. No longer nurtured by such bohemias, intellectuals moved to college campuses and addressed other scholarly colleagues through the specialized journals and rhetorics of their disciplines.

9. In 1950 coal accounted for 47 percent of net electric power generation in the United States, with none being provided by nuclear power. By 1970, coal accounted for 46 percent and nuclear 1.4 percent. Twenty years later, coal's share had risen to almost 56 percent and nuclear's to almost 21 percent (U.S. Department of Energy 1990)

Billion KWh Net Generation, 1950–1990

Year	Coal	Petrol	Nat Gas	Nuclear	Hydro	Other	Total
1950	155	34	45	0	96	—	329
1960	403	48	158	1	146	—	756
1970	704	184	373	22	248	1	1532
1980	1162	246	346	251	276	6	2286
1990	1557	117	263	577	280	11	2805

10. Beginning in 1969, John Friedmann has sought to bring theoretical coherence to the idea of "radical planning," first by introducing the notion of "system transformation" (1969), then by discussing social learning and transactional planning (1973), and finally by arguing that planners should actively help to "emancipate" humanity from social oppression (1987). In his view, the thrust for transformative change must come from "below," the struggle for liberation must be a form of self-liberation, and people have to take hold of their own lives in their own communities rather than leave things to the State. Radical planners in his view should work with such communities. See also Grabow and Heskin 1973 and Mazziotti 1974.

11. The inability of modernist planning to deliver on its promise was perhaps best exemplified by President Carter's unsuccessful 1977–79 effort to develop a coherent national energy policy. Energy experts working in isolation developed a comprehensive energy plan, but Congress rather quickly threw out its major elements (Chubb 1983; Katz 1984).

12. As Elizabeth Wilson (1991, 150–51) puts it, in the late nineteenth and early twentieth centuries "utopians, planners and architects believed that the only solution was to scrap the existing unplanned, irrational cities and build new, planned ones. Today, by contrast, planning, planners and architects are blamed for having *caused* the current state of our cities by their overweening interference." Wilson goes on to suggest, quite rightly I think, that "critical of the disasters of utopian planning, we are in danger of forgetting that the unplanned city still *is* planned, equally undemocratically, by big business and the multinational corporation" (152).

Chapter Two

13. In *Nature's Metropolis,* William Cronon (1991) argues that it is a mistake to think of the urban and rural landscapes as being two separate places. In his view, a city such as Chicago was formed out of nature's materials (coal, for example, for producing electric power), while that city's rural hinterland was transformed from deciduous forests into board feet of lumber and from

tall grass prairie to fields of grain. Thus two seemingly different landscapes are really one: "they created each other, they transformed each other's environments and economies, and they now depend on each other for their very survival. To see them separately is to misunderstand where they came from and where they might go in the future. Worse, to ignore the nearly infinite ways they affect one another is to miss our moral responsibility for the ways they shape each other's landscapes and alter the lives of people and organisms within their bounds" (384–85).

14. Here I refer primarily to Habermas's (1987) effort to reconstruct the dialectic of the Enlightenment in terms of "communicative reason"; to Gadamer's (1991) call for dialogue oriented toward a "fusion of horizons"; to Rorty's (1979 and 1989) pragmatic conception of rationality as civility; to Lyotard's (1984) reference to the breakdown of all grand narratives; to Geertz's (1983) emphasis on the locality and contextuality of knowledge; to Foucault's interpretation of professionalized "discursive formations" as will-to-power (see Sarup 1989 and Boyne 1990); and to Derrida's claim that scientific texts can be "deconstructed" as works of literature (see Sarup 1989 and Boyne 1990).

15. Toulmin (1990) provides a complementary view when he argues that nineteenth century philosophers held that the validity and truth of "rational" arguments was independent of who presented them, to whom, or in what context. To the contrary, Toulmin argues that "the dream of finding a scratch line, to serve as a starting point for any 'rational' philosophy, is unfulfillable. *There is no scratch*. . . . All we can be called upon to do is to take a start *from where we are, at the time we are there*" (178–79). We need, he argues, to broaden "the 'modern' focus on the written, the universal, the general, and the timeless . . . to include once again the oral, the particular, the local, and the timely" (186).

16. As should be clear from the preceding chapter, not all planning is performed by professional planners, and some professional planners do not work at activities labeled as planning (Dalton 1989). For the purposes of this chapter, I define the term "planner" to include any person or organization that is engaged in a conscious social effort to guide change.

17. Scientific planners do not, of course, claim to be engaged in persuasion. Imagining themselves to be standing on the central plateau where Truth speaks for itself, they portray themselves as "anti-rhetoricians" who are guided by logic, not rhetoric. John Locke (cited in Simons 1990, 1) nicely expresses the antirhetorical point of view:

> If we would speak of things as they are, we must allow that all the art of rhetoric, besides order and clearness . . . are for nothing else but to insinuate wrong ideas, move the passions, and thereby mislead the judgement; and so indeed are perfect cheats; and therefore . . . they are certainly, in all discourses

that pretend to inform or instruct, wholly to be avoided; and where truth and knowledge are concerned, cannot but be thought a great fault, either of the language or person that makes use of them.

18. The problem is, at least within the modernist context, that this conception of planning as a moral activity draws planning into *emotivism.* According to this philosophical doctrine, moral judgments can be nothing but expressions of preference, attitudes or feelings. Even so, each moral claimant believes that his or her judgments somehow represent impersonal criteria and ultimate principles. Consequently, disputes between conflicting moral arguments cannot be resolved rationally (MacIntyre 1981). Compelled by media technologies to simplify, contemporary advocates indulge in sloganeering and other rhetorical extremes that intensify into "cultural wars" (see Hunter 1991). Though propounding their claims with vigor, they cannot establish the truth of their claims through scientific or ethically neutral reasoning. Talking past one another, and denying any validity to their opponents' claims, they seek to impose their cultural and moral ethos upon others through political means. Given this incommensurability of moral claims, it is easy to see how one planner's utopia might prove to be another person's "cacotopia," and that implemented plans might leave some residents feeling like Franz Kafka's Gregor in *The Metamorphosis* or like the subject of Edvard Munch's famous painting *The Scream.*

19. Machiavelli provides seminal guidance to the modern politician: "A man who tries to be good all the time," he writes, "is bound to come to ruin among the great number who are not good. Hence a prince who wants to keep his post must learn how not to be good, and use that knowledge, or refrain from using it, as necessity requires" (1977, 44). Furthermore, the prince "must always look to the end. Let a prince, therefore, win victories and uphold his state; his methods will always be considered worthy, and everyone will praise them, because the masses are always impressed by the superficial appearance of things, and by the outcome of an enterprise" (51).

20. Consider the rhetorics of science and advocacy as they pertain to the nuclear power industry in the late 1970s and early 1980s. Weart (1988) reports that "Some people saw the antinuclear groups as brave coalitions of potential victims, while officials were dangerous men with ice water for blood. . . . Meanwhile other people identified officials with benevolent forces of reason, opposed by an ignorant and rebellious mob" (353). He further observes that these two groups presented their arguments to two very different audiences: "Most of the articles explaining that nuclear industry was safe and beneficial were printed in the recondite technical journals of physics, health physics, and engineering. On the paperback racks in bookstores, most of the books about nuclear power were openly hostile" (362).

21. By drawing this analogy to fiction, I do not mean to suggest that plan-

ners simply "make up" stories. Rather, I mean to stress the importance of thinking of planners as authors who actively construct views of events.

22. For planners, that ordinarily means writing as either an "objective third-person author" or as a "peripheral first-person narrator." In the first case, the author describes only that which might be observed by a person; the author is impersonal but not omniscient. In the latter, the author "speaks" in the first person through a narrator who has all the limitations of an ordinary human being; such a narrator can describe, interpret, explain, and predict, but only in the fallible way available to humans.

23. In a similar vein, Hunter (1991) urges us to establish a context of public discourse that can sustain a genuine and peaceable pluralism. For Hunter, that context would contain four elements: (1) changing the environment of public discourse to promote genuine debate through direct and immediate exchanges; (2) having all factions reject the impulse to remain publicly quiescent; (3) recognizing the "sacred" within different moral communities; and (4) recognizing the inherent weaknesses, even dangers, in each moral community's own commitments.

Chapter Three

24. In a franchise agreement, a municipal authority normally grants an investor-owned utility the right to use public property (e.g., streets) needed for placing electrical transmission and distribution equipment. In return, the utility usually pays the municipality a small fee or provides free electric service. See Romo 1989.

25. Though later portions of this book delve more deeply into the regulatory bargain's implications for planning and rate making, the book does not seek to act as an advanced textbook for regulators. For that kind of detail, readers should turn to Phillips 1993 or Bonbright, Danielsen, and Kamerschen 1988.

26. The electric power industry measures power (work divided by time) in terms of watts (W). One thousand watts equals one kilowatt (kW), and 1,000 kW equals one megawatt (MW). It measures energy (work, or the product of power and time) in terms of watt-hours (WH). One thousand WH equals one kilowatt-hour (kWh). Capacity (the ability to produce a given output of electricity at an instant in time) is a measure of power, whereas output (the amount of electricity produced in a period of time) is a measure of work. A nuclear generating unit with a capacity of 1,000 MW can provide enough electric energy to serve a city of 400 thousand people. If it operates at 80 percent capacity for an entire year, it will generate slightly over 7,000,000 MWH per year.

27. In *Federal Power Commission v. Hope Natural Gas Co.* (1944), the U.S.

Supreme Court held that "Under the statutory standard of 'just and reasonable' it is the result reached not the method employed which is controlling. . . . It is not theory but the impact of the rate order which counts. If the total effect of the rate order cannot be said to be unjust and unreasonable, judicial inquiry under the Act is at an end. . . . Moreover, the Commission's order does not become suspect by reason of the fact that it is challenged. It is the product of expert judgment which carries a presumption of validity" (cited in Phillips 1993, 329).

28. The economic version of this argument was first advanced by Averch and Johnson's "Behavior of the Firm under Regulatory Constraint" in *American Economic Review* (1962). For a review of that article and subsequent commentary on it, see Phillips 1993, 892–93.

Chapter Four

29. Classical rhetoricians might characterize this steady drumbeat of rate increases (the feeling of which I have tried to capture both in the style and structure of this chapter) as *accumulatio*; that is, a series of statements that heap praise or blame in order to emphasize a point already made (see Quinn 1982 and Lanham 1991).

30. For more detailed insight into the collapse of the nuclear power industry nationwide, see Cook 1985, Rudolph and Ridley 1986, Campbell 1988, and Morone and Woodhouse 1989.

31. Opponents of the Seabrook and Shoreham nuclear power plants, in New Hampshire and New York respectively, successfully used the NRC's regulations to delay efforts to develop emergency evacuation plans for their localities. In late 1988, President Reagan signed an order allowing the Federal Emergency Management Agency to devise and carry out evacuation plans without the cooperation of local governments in order to override local opposition to nuclear power plants. Advocates of nuclear power were pleased by the President's action; opponents were dismayed. For a fascinating if one-sided story of the effort to build the Shoreham plant on Long Island and of local government opposition to that effort, see Grossman 1986. Public fears forced the Long Island Lighting Company to abandon the $5.5 billion plant, even though it had been built and fully licensed by the NRC. It sold the plant to the state for $1. Peter A. Bradford, chair of the New York Public Service Commission, said, "What has happened on Long Island in the last three years makes clear beyond the possibility of argument that there are alternatives to nuclear fuel. The fact that it was possible to replace Shoreham at less cost to the consumer than operating it speaks volumes about whether nuclear power is competitive right now" (Lyall 1991).

32. The genealogy of least-cost planning can be traced through the Energy

Policy Project of the Ford Foundation (1974), Amory Lovins (1976), Roger Sant (1979), the Solar Energy Research Institute (1981), Cavanagh (1986), and others.

33. For further insight into alternative forecasting models, see Committee on Energy and Commerce 1984 and Fiddyment 1986.

Chapter Five

34. The three cases concerned (1) whether Edison should refund approximately $70 million to ratepayers for inefficient use of the LaSalle 1 unit in 1982; (2) whether Edison's rates should be reduced in light of federal tax reductions scheduled to take effect in 1987; and (3) whether Edison's rate of return on investment should be reduced to reflect recent trends in the cost of borrowing money.

35. The prudency standard refers to managerial planning and decision making; ratepayers should not have to pay for the costs of imprudent investments. The used and useful standard refers to plants that are actually being used to generate or transmit useful electric power. A utility plant can fail the used and useful test even if it was prudently incurred.

36. Edison served the City pursuant to a franchise that was due to expire in late 1990. Disturbed by Edison's rate increases, Washington began exploring options to remaining on Com Ed's system in late 1985. For details, see chapters 7 through 9.

37. Other interpretations could, of course, be offered. One might, for example, focus narrowly on the negotiation process and, drawing on the relevant literature (e.g., Susskind and Cruikshank 1987), argue that Edison's negotiating strategy was faulty. Alternatively, one could draw considerable interpretive insight from poststructuralists such as Foucault and Derrida (see Boyne 1990, and Sarup 1989) and critical theorists such as Habermas (1987) and Forester (1989). Foucault might claim that the story I have recounted should turn less on duplicity and more on how Edison, its regulators, and its consumer opponents have created a professionalized "discursive formation" that seeks to impose its order on others. Derrida, on the other hand, might argue that each planning text noted in the story refers to no "reality" other than itself and other texts. Edison's "negotiated settlement" was not a "settlement" except on its own terms, and the merits of that "settlement" could be measured only by relying on contestable forecasts and contestable interpretations of the Public Utilities Act. Habermas and Forester might interpret my story as revealing the ways in which Com Ed's claims (e.g., that they had "negotiated" a "settlement," that their proposal was the "best deal" for consumers, and that all consumers had to do was "pick up the telephone" if they wanted to influence the ICC staff) reproduced a world of big organizations in which "opportu-

nity" seems equal but some players have far greater resources than others (Forester 1990).

38. To make one's final offer first (e.g., $660 million and no less) is a negotiation blunder of the first magnitude, for it leads (and led) to charges of negotiating in bad faith if one backs off later. See Susskind and Cruikshank 1987.

Chapter Six

39. The larger story is more complicated. For example, the company's application for a traditional rate increase also would have altered the differential between summer and nonsummer rates for residential customers. In the late 1970s, the ICC had mandated higher rates in the summer as a way of reducing growth in peak demand and hence restraining the need for new capacity. But consumers were now complaining about having to pay large electricity bills in the summer. CUB asked the Illinois Appellate Court to grant an immediate 25 percent summer rate cut for customers. On April 27, 1988, the ICC ordered Edison to alter residential rates so as to reduce summer rates by 25 percent. Edison believed that the order would have the effect of reducing the company's revenues by $142 million in 1988, so it appealed the order. In May, the Appellate Court blocked implementation of the Commission's order. A month later, the company filed (with the Commission's approval) new rates that would, in its judgment, eliminate that lack of "revenue neutrality" and reduce summer rates by only 13 percent. Several intervenors appealed the Commission's order. In December 1988, the Appellate Court declared the Commission's June decision void and removed the stay of the April order. The Attorney General and the Cook County State's Attorney asked the ICC to order refunds amounting to roughly $150 million. Edison, on the other hand, argued that its should refund no more than $5.7 million. The case was back before the ICC at the end of 1989 (Com Ed 1989).

40. According to the modernist view of planning, Arthur Young's auditors sought to *measure* objectively and precisely the unreasonable portion of Edison's powerplant construction expenditures. I would argue that their audit sought both to quantify costs and to legitimate the ICC's impending course of action. Thus I would interpret the audit as a trope in persuasive storytelling. For an insightful discussion of auditing as a hermeneutical practice, see Francis 1994.

41. One might also wonder who will pay for any nuclear accident that might occur at one of Edison's facilities. The company is a member of Nuclear Mutual Limited (NML), which was created after the Three Mile Island accident to provide insurance coverage against property damage to members' nuclear generating facilities. The company would be subject to a maximum

assessment of roughly $162 million in any one year in the event that losses exceeded accumulated reserve funds. Edison is also a member of Nuclear Electric Insurance Limited (NEIL), which provides insurance coverage against the cost of replacement power obtained during certain prolonged accidental outages of nuclear units and coverage for property losses in excess of $500 million occurring at nuclear stations. The company could be subject to maximum assessments, during any one year, of approximately $36 and $52 million in the event losses exceeded accumulated reserve funds under the replacement power and property damage coverages respectively. In addition, the Price-Anderson Act establishes a maximum liability limit in the event of a nuclear accident: the company would be subject to a maximum assessment of $787.5 million in the event of an incident, limited to a maximum of $125 million in any one year (Com Ed 1988, 40–41).

Chapter Seven

42. For a definition of franchise agreement, see note 24, chapter 3.

43. "Wheeling" refers to the use of the electric power transmission facilities of one system to transmit power produced by other entities.

44. Included on the Task Force were representatives of the Center for Neighborhood Technology (CNT), the Labor Coalition on Public Utilities, the Illinois Public Action Council (IPAC), CUB, and sixteen other agencies and organizations.

45. Fremon (1988) reported that Edison spent $630,000 for advertisements in just three newspapers (the *Tribune,* the *Sun-Times,* and the *Defender*) during the last two months of 1987.

46. Edison's unflattering portrait of the City Council in this flyer stands in interesting juxtaposition to the portrait presented in Hogan's (1986) friendly history of the utility. There Hogan implies that Com Ed executives expect politicians to bluster and posture in public (to obtain public support) but wheel and deal in private, all the time trying to enhance their personal wealth and power.

47. Headed by KEA, the team included Ralph Cavanagh and his associates at the Natural Resources Defense Council, MSB Energy Associates, and M. R. Beal and Company.

48. Members of the CEOC included CUB, CNT, IPAC, National Peoples Action, Nuclear Energy Information Service (NEIS), the South Austin Coalition Community Council, Organization of the Northeast, the Logan Square Neighborhood Association, and seven other groups.

49. "Cogeneration" is a technical process which involves the simultaneous generation of electric power and usable heat. "Least-cost planning" refers to a planning process that places traditional supply-side alternatives (e.g., nuclear plants) on an equal footing with demand-side alternatives (e.g., fluorescent

light bulbs that use energy more efficiently). See Markowitz and Hirsh 1987 and Cavanagh 1986.

50. Acting as a subcontractor to Calmar, the Center for Neighborhood Technology would investigate the economic impact of Edison's rate increases.

Chapter Eight

51. The Task Force consisted of thirty-six private citizens of varied backgrounds—including business, professional, labor, and community representatives (MTFE 1989a).

52. A reader might wonder about the methods used to select and research this particular case. As the first portions of the chapter should have made clear, the first step was to learn about the larger narratives of which the case is a part. Details about those narratives can be found in earlier chapters and in Throgmorton 1992 and Throgmorton and Fisher 1993. The flow of those narratives strongly suggested that the CACI's survey marked a potentially critical turning point in the controversy. Subsequent conversations with CEOC people and Chicago Planning Department staff led me to learn that the Department had videotaped a Mayor's Task Force meeting in which Michael McKeon discussed the CACI's survey. (The meeting was also shown on the Chicago municipal information television channel, WCTV.) I obtained a copy of the videotape and watched it four times. Struck by its relevance to the rhetoric of planning, I asked a graduate student, Michelle Javornik, to prepare a verbatim transcript of the meeting, omitting such verbal ticks as "ah" and "uh." Drawing on my close viewing of the videotaping, I added comments about physical gestures where appropriate. To save space, I also deleted redundant language (marked by . . .) that distinguished the oral from the written word. To test my own viewing for potential bias, I also showed the tape to four colleagues and to graduate students in one of my courses. Their responses were, in all cases except one, consistent with mine. One major drawback to the tape is that it rarely shows members of the audience reacting to a speaker's comments and gestures. See Mayor's Task Force on Energy 1989b.

53. See note 49, chapter 7, for a definition of cogeneration. A large-scale shift to cogeneration by businesses could have a dramatic effect on Edison's position as the sole provider of electric power in the Chicago area.

54. Mitchell's and McKeon's responses seemed to satisfy the Task Force members, for they turned to a new topic. But note that a *Commerce* magazine editorial (Anonymous 1988) leaves it unclear whether there were additional questions, saying only that "Selected questions and findings of the CACI survey include the following."

55. Some readers might argue that the story recounted in the present chapter is not really about planners except in the broadest sense. I would respond that Michael McKeon surely is not a planner in the narrow sense but that

survey research is a vital part of planning and that his survey acted as an important trope in the City's effort to explore options. A great deal can be learned by observing practitioners, regardless of whether they are formally trained in planning. My sense is that most planning schools tend to teach their students how to do survey research, then have them read an article or two about how to be a survey researcher. That is not a good balance. Students cannot understand survey research and other planning tropes without being aware of the diverse contexts in which they will be used and defended.

56. Babbie (1973, 362) reflects this ethos when he insists that "each scientist operates under a normative obligation to share his findings with the scientific community, which means sharing them with nonscientists as well." For Babbie that means not just sharing "findings" but providing enough information to enable a reader to replicate the entire survey independently.

Chapter Nine

57. Reading an earlier version of the present book, one reviewer observed that Edison officials might have been of two minds at this point in time, with both types worried about the company's increasing vulnerability. One type ("corporate true believers") may really have believed that the company's authority derived from its rational expertise. They would have dismissed rhetoric as mere words that add gloss to the important stuff. The second type ("cynical officials") may have been quite conscious of using language to seduce, manipulate, and defend their deployment of power.

58. Included in this group were spokespersons for CNT, the Labor Coalition, CUB, CEOC, the Northwest Community Organization, South Austin, the Forty-sixth Ward Fair Share, NEIS, and National People's Action.

59. A spokesperson for the CEOC noted that "We spend at Whitney Young $247 just about per child per year for electricity, and we're only spending $37 on books per child per year. And I call that a crime. That's obscene" (Mayor's Task Force on Energy 1989b, May 2).

60. Included among the Working Group's technical advisors were Coopers and Lybrand (an accounting firm), Smith Barney Harris Uphan and Company (a financial consulting firm), and Mayer Brown and Platt (a law firm). Helman, cochair of the law firm's management committee, had negotiated several large corporate takeovers in the past. "I'm the head of this very large law firm, and I've done a lot of very large transactions," he said; "I take this very seriously" (Dold 1990).

61. This fifty-nine page report drew little public attention. Charles Komanoff, who authored the report, concluded that acquistion of some or all of Edison facilities within the city and establishment of a City-owned electric utility system was "a viable option" but that the potential for clear-cut rate

reductions indicated in the 1987 Beck report "appears to have diminished," largely "because of unfavorable changes in the federal tax code" (1). Komanoff further concluded that improved end-use efficiency offers "tremendous potential for reducing electricity costs and bills" but that "large-scale investment by the City's electricity provider will be required to exploit the full potential for cost-effective improvements in end-use efficiency" (largely because "market barriers" would otherwise block the way) (1–2). "In short," he wrote, "firm and skillful negotiation of a new franchise with Edison may allow the City to obtain most or all of the benefits that might be available through exercising the option to acquire and operate Edison facilities through a municipal utility, without the attendant risks. At the same time, the City retains a viable acquisition option 'in its back pocket' if Edison refuses to make the appropriate changes and concessions" (2).

62. One latent issue in the Chicago franchise case was whether residents of one town should have to pay higher bills to offset franchise fees that their utility companies must pay to do business in a nearby city; i.e., whether utilities should pass off local franchise fees to customers throughout their service territories. In March 1990, in a case involving a small central Illinois water company, the ICC declared that the water utility could recover the cost of Champaign and Urbana franchise fees only from Champaign and Urbana residents. It agreed with an ICC staff statement: "'Franchise fees are a hidden tax imposed by the cities and, as such, should be recovered from customers located in the area benefiting from the tax'" (Karwath 1990a).

63. The full consultant report had not been made available to the public because Edison had required the Daley adminstration to sign a "confidentiality agreement" before sharing information with the City about the West Side blackouts and maintenance of the utility's transmission and distribution system (Ylisela and Capitanini 1991).

64. Edison indicated that the downtown part of Chicago needed more power and that the company needed to extend a 345 kilovolt transmission line from Sixty-third to Twenty-third Streets. By the middle of 1990, some residents and institutions in that part of the city had started opposing Edison's plan. Researchers at the Illinois Institute of Technology (which is located next to the planned route for the transmission line) expressed concern about the safety of the electric and magnetic fields created by the lines. The Chicago Housing Authority (CHA), which managed a vast array of public housing in the area, and several south side organizations also opposed it. "There are 30,000 public housing residents who live within a half-block to a block of the spot where the power lines are going to be erected," said a CHA spokewoman. "Our concerns are for their health and safety." "Everybody is sort of shocked," another person said. According to Edison officials, there was no evidence that

the magnetic fields posed any undue risk, whereas burying the line underground would increase the project cost from $43 million to $83 million (Brotman 1990).

65. From late October through mid-November, Edison ran a series of large advertisements opposite the *Chicago Tribune*'s editorial page. One of them consisted of a bordered blank rectangle, below which appeared the words "Burn this ad." Many observers concluded that this series of advertisements simply paid the *Chicago Tribune* for its vigorous and unwavering editorial support of Edison over the years. A person named Greg Anderson was one. His letter of December 14, 1991, to the *Tribune*'s editors said that "One-fourth of your Op-Ed was filled by an ad for Commonwealth Edison on the very day that the utility's franchise agreement was voted on by the City Council Energy Committee. . . . Is that was you call the free exchange of ideas? Or are we supposed to pay to appear on your Op-Ed page? . . . [T]o run an ad on your Op-Ed page implies that advertising is what forms your opinions" (Anderson 1991). Carolyn Ross was another. Her letter to the *Tribune*'s editors noted that "For several days recently, they [Edison] have run full-page ads in the Tribune . . . According to the Tribune advertising department, this ad costs $40,876 [for full-run] the first time it appears and $15,080 per day thereafter" (Ross 1991).

66. The *Tribune* also reported (Kass 1991d) that some aldermen saw the franchise as an opportunity to obtain a few more patronage jobs and to obtain favorable ward boundaries in the redistricting that was currently underway. (The redistricting had to be passed by the full council before December 1 to avoid a citywide referendum.)

Chapter Ten

67. On February 23, 1990, the ICC ordered Edison to refund $5.7 million plus interest to residential customers for excess revenues stemming from the ICC's 1988 summer/nonsummer rate decision and the June 1989 Supreme Court ruling overturning that decision. Claiming that the refund should have been $177 million, consumer groups and Attorney General Hartigan appealed to the Appellate Court. In July 1991, the Appellate Court reversed the ICC's 1990 decision and instructed the Commission to use a different methodology when reconsidering the case (Com Ed Annual Report 1992).

68. One of the examiners, Therese Freund, was involved in both proceedings. Furthermore she had been involved in virtually every major Edison rate case in the 1980s. Consumer groups repeatedly complained that her draft orders consistently favored Edison and twice tried to have her removed from Edison cases (Karwath 1990f).

69. Other intervenors included the Office of Public Counsel, the Cook County State's Attorney, the City of Chicago, the Labor Coalition on Public

Utilities, Community Action for Fair Utility Practice, and Illinois Industrial Energy Consumers.

70. Recall that in June 1987 the Supreme Court had ordered the ICC to reconsider its decision granting Edison a rate increase for Byron 1. Recall too that two years later the ICC had ordered Edison to remove $200 million of unreasonable construction costs from its rate base and to refund approximately $190 million to its customers. And the Circuit Court of Cook County had ruled late in 1989 that the refund should include 9 percent interest on the money to be refunded, rather than 5 percent as the Commission originally ordered. In the Circuit Court's view, the refund should total about $248 million.

Edison and several intervenors appealed the Circuit Court's decision. On June 7, 1991, an Appellate Court decided that the ICC, rather than the Circuit Court, was in the best position to determine how large the refund should be. It remanded the proceeding to the ICC, instructing the Commission to determine whether the refund should be based on revenue projections the company had made in 1985 or on the revenues it actually collected under the rate increase. The Appellate Court also instructed Edison to "promptly implement" the refunds for April 1986 through December 1988, with 5 percent interest. Edison and several other intervenors appealed the Court's ruling to the Supreme Court, which in turn heard oral arguments in late June 1991. In the second quarter of 1990, Com Ed wrote off $208 million, or $0.98 per common share, in response to the Appellate Court's decision (Com Ed 1992).

Edison had to write off even more after the Supreme Court overturned part of the Appellate Court's decision on April 16, 1992. The Supreme Court agreed with the Commission that Edison should absorb the $291 million in disallowances for Byron 1, said that the refund should be based on Edison's actual revenues rather than the ones it forecasted back in 1985, and concluded that Edison should pay 9 percent interest on the refunds. So the Supreme Court sent the case back to the Circuit Court with instructions to implement the refunds. It also sent two unresolved issues back to the ICC: how to deal with indirect costs and how refunds for the period after December 31, 1988, should be calculated. As a result of this decision, Com Ed would have to refund $225 to $260 million to its customers and write off $65 million, or 18¢ per common share, during the first quarter of 1992. Edison was not pleased: "We're being clobbered by the courts, having money taken away from us," said one of its spokesmen. "I don't know how we can deliver the service that's expected" (Grady 1992).

71. Chapter 1 emphasizes that early advocates of electrification argued that electricity would increase man's control over nature. Though those advocates clearly meant increasing *man's* control, many early feminists and domestic scientists (Ellen Richards, for example) conceived of electricity as "a drudge

and willing slave" (Nye 1990, 247) and of electric appliances as replacements for servants. In their view, the job of being housewife and mother should focus primarily on child care and nurturing in the home, and the complex inner working of appliances and the electric power systems that activated them should be left to external experts, presumably male. Thus in one sense electrification has had regressive effects on women. By the 1980s, moreover, it was not unusual to find some ecofeminists arguing that there are important connections between the oppression of women and the oppression of nature. Warren (1987), for example, argues that these oppressions are sanctioned by a dominant "patriarchal conceptual framework" which is characterized by (1) "value-hierarchical thinking," which gives rise to (2) a "logic of domination," which in turn generates (3) "normative dualisms" (e.g., rational/emotional, cultural/natural, scientific/political) wherein one pole is assigned higher value than the other. Still, Nye reports that electrification also helped to blur sex roles by making it easier for women to enter the work force and for men to live independently as bachelors.

Chapter Eleven

72. "Gridlock" is another contemporary (if more bureaucratic) way of characterizing this frozen embrace. The smooth flow of planning and policy making is blocked at every turn; no one firm, organization, or group has sufficient power to impose its will on others; many have enough power to block initiatives that they oppose.

73. Edward Said adopts a similar approach in his insightful recent book, *Culture and Imperialism.* Describing his approach as "contrapuntal reading," he writes that "In juxtaposing experiences with each other, in letting them play off each other, it is my interpretive political aim (in the broadest sense) to make concurrent those views and experiences that are ideologically and culturally closed to each other and that attempt to distance or suppress other views and experiences (1993, 32–33). He too believes that it is important to support minority and "suppressed" voices while also retaining vigilance and self-criticism: "there is an inherent danger," he writes, "to oppositional effort becoming institutionalized, marginality turning into separatism, and resistance hardening into dogma. . . . [Hence,] there is always a need to keep community before coercion, criticism before mere solidarity, and vigilance ahead of assent" (54).

References

Abu-Lughod, Janet L. 1991. *Changing Cities.* New York: HarperCollins Publishers.

Alexander, Ernest. 1984. After rationality, what? *Journal of the American Planning Association* 50: 62–68.

Alterman, Rachelle, and Duncan MacRae, Jr. 1983. Planning and policy analysis. *Journal of the American Planning Association* 49: 200–15.

Anderson, D. D. 1981. *Regulatory Policies and Electric Utilities: A Case Study in Political Economy.* Boston: Auburn House.

Anderson, Greg. 1991. Paid views. Letter to the editor of the *Chicago Tribune* (December 14).

Anonymous. 1988. Strong opposition to municipalized electric service. *Commerce Magazine* (January).

Anonymous. 1990. Hartigan vows to fire ICC chief. *Chicago Tribune* (October 22).

Anspaugh, Lynn R., Robert J. Catlin, and Marvin Goldman. 1988. The global impact of the Chernobyl reactor accident. *Science* 242: 1513–19.

Arndt, Michael. 1987a. Edison files for rate increase: Bargaining with citizen group fails. *Chicago Tribune* (February 4).

———. 1987b. ICC gives Edison a reprieve on Braidwood case. *Chicago Tribune* (February 19).

———. 1987c. ICC orders speed-up of Edison rate plan hearings. *Chicago Tribune* (March 19).

———. 1987d. Edison puts high price on facilities: Cost to city estimated at $2.8 billion. *Chicago Tribune* (November 9).

Arndt, Michael, and Jean Davidson. 1987. Edison bid out of line, report says. *Chicago Tribune* (April 28).

Averch, Harvey, and Leland Johnson. 1962. Behavior of the firm under regulatory constraint. *American Economic Review* 52: 1053.

Babbie, Earl R. 1973. *Survey Research Methods.* Beaumont, CA: Wadsworth Publishing.

Barnett, Jonathan. 1986. *The Elusive City: Five Centuries of Design, Ambition and Miscalculation.* New York: Harper & Row.

Baum, Howell S. 1983. *Planners and Public Expectations.* Cambridge, MA: Schenkman Publishing Co.

Bauman, John F. 1987. *Public Housing, Race, and Renewal: Urban Planning in Philadelphia, 1920–1974.* Philadelphia: Temple University Press.

Bazerman. C. 1988. *Shaping Written Knowledge.* Madison: University of Wisconsin Press.

Beauregard, Robert A. 1989. Between modernity and postmodernity: the ambiguous position of U.S. planning. *Environment and Planning D: Society and Space* 7: 381–95.

———. 1990. Bringing the city back in. *Journal of the American Planning Association* 56: 210–15.

———. 1991. Without a net: Modernist planning and the postmodern abyss. *Journal of Planning Education and Research* 10: 189–93.

Beck, R. W., and Associates. 1987. *Electric Supply Options Study.* Prepared for the City of Chicago, Department of Planning. Indianapolis: R. W. Beck and Associates.

Bennett, Larry. 1988. Harold Washington's Chicago: Placing a progressive city administration in context. *Social Policy* 19, no. 2 (fall): 22–28.

Benveniste, Guy. 1989. *Mastering the Politics of Planning.* San Francisco: Jossey Bass.

Berman, Marshall. 1982. *All That Is Solid Melts into Air: The Experience of Modernity.* New York: Penguin Books.

Berman, Paul, ed. 1992. *Debating P.C.: The Controversy over Political Correctness on College Campuses.* New York: Laurel.

Bernstein, Richard J. 1983. *Beyond Objectivism and Relativism: Science, Hermeneutics, and Praxis.* Philadelphia: University of Pennsylvania Press.

Biddle, Fred Marc. 1985a. City may unplug Edison, run own utility. *Chicago Tribune* (October 10).

———. 1985b. Edison may leave if city pulls plug. *Chicago Tribune* (October 14).

———. 1988. ICC to study rate hike for Edison. *Chicago Tribune* (August 2).

Biddle, Fred Marc, and Steve Johnson. 1987. Hopes dim for Edison compromise. *Chicago Tribune* (January 23).

Black, Alan. 1990. The Chicago area transportation study: A case study of rational planning. *Journal of Planning Education and Research* 10: 27–37.

Bluestone, Barry, and Bennett Harrison. 1982. *The Deindustrialization of America: Plant Closings, Community Abandonment, and the Dismantling of Basic Industry.* New York: Basic Books.

Bonbright, James C., Albert L. Danielsen, and David R. Kamerschen. 1988.

Principles of Public Utility Rates. 2d ed. Arlington, VA: Public Utilities Reports, Inc.

Boyer, M. Christine. 1983. *Dreaming the Rational City: The Myth of American City Planning.* Cambridge, MA: MIT Press.

Boyne, Roy. 1990. *Foucault and Derrida.* London: Unwin Hyman.

Broad, William J. 1987. Rise in retarded children predicted from Chernobyl. *New York Times* (February 16).

Brooks, Michael P. 1988. Four critical junctures in the history of the urban planning profession: An exercise in hindsight. *Journal of the American Planning Association* 54: 241–48.

Brotman, Barbara. 1990. Power line generates controversy. *Chicago Tribune* (June 1).

Bruder, Stanley. 1990. *Visionaries and Planners: The Garden City Movement and the Modern Community.* New York: Oxford University Press.

Bukro, Casey. 1986. Critic of nuclear power urges Edison to phase out reactors. *Chicago Tribune* (June 10).

Bupp, Irvin C., and Jean-Claude Derian. 1978, 1981. *The Failed Promise of Nuclear Power: The Story of Light Water.* New York: Basic Books, Inc.

Burchell, Robert W., ed. 1988. Symposium on planning, power and politics. *Society* 26: 4–42.

Burchell, Robert W., and James W. Hughes. 1980. Planning theory in the 1980's—A search for future directions. In *Planning Theory in the 1980's,* edited by Robert W. Burchell and James W. Hughes. Piscataway, NJ: CUPR Press.

Burleigh, Nina. 1989. Power plays: The public relations battle between community organizers and Commonwealth Edison. *Neighborhood News* (January 13).

Burroway, J. 1987. *Writing Fiction: A Guide to Narrative Craft.* 2d ed. Glenview, IL: Scott, Foresman and Company.

Business and Professional People for the Public Interest (BPI). 1987. Brief of Business and Professional People for the Public Interest in the Matter of Commonwealth Edison Company, People of the State of Illinois, Citizens Utility Board, 87-0043, 87-0044, 87-0057, and 87-0096 (June 1).

Business and Professional People for the Public Interest and Citizens Utility Board (BPI and CUB). 1989a. Brief of Business and Professional People for the Public Interest and Citizens Utility Board in the Matter of Business and Professional People for the Public Interest, et al., v. Illinois Commerce Commission and Commonwealth Edison Company, 68100, 68246, 68247, 68306, and 68355 (Consolidated) (March 1).

———. 1989b. Brief of Business and Professional People for the Public Interest and Citizens Utility Board in the Matter of Business and Professional

People for the Public Interest, et al., v. Illinois Commerce Commission and Commonwealth Edison Company, 68100, 68246, 68247, 68306, and 68355 (Consolidated) (April 25).

Campbell, John L. 1988. *Collapse of an Industry: Nuclear Power and the Contradictions of U.S. Policy.* Ithaca, NY: Cornell University Press.

Camper, John, and Dave Schneidman. 1985. Edison rate hike fuels opposition. *Chicago Tribune* (October 27).

Camper, John, and Mark Eissman. 1987. Edison rate plan wins round. *Chicago Tribune* (June 13).

Caro, Robert A. 1974. *The Power Broker: Robert Moses and the Fall of New York.* New York: Vintage.

Carson, Rachel. 1962. *Silent Spring.* Boston: Houghton Mifflin.

Carter, Luther J. 1987. *Nuclear Imperatives and Public Trust: Dealing with Radioactive Waste.* Washington, DC: Resources for the Future.

Castells, Manuel. 1989. *The Informational City.* New York: Basil Blackwell.

Cavanagh, Ralph C. 1986. Least-cost planning imperatives for electric utilities and their regulators. *The Harvard Environmental Law Review* 10: 299–344.

Chicago Electric Options Campaign (CEOC). 1989. Notes for neighborhood presentations (April 14).

Chicago Tribune. 1985. A brownout in City Hall (October 1).

———. 1987. Don't unplug Edison in Chicago (October 30).

———. 1988a. Chicago should study power options (January 9).

———. 1988b. End the war over Edison's rates (May 1).

———. 1988c. A good move to end Edison rate fight (June 15).

———. 1989. Running electricity through City Hall (December 13).

———. 1990a. Dignity amidst disaster on West Side (August 1).

———. 1990b. Daley dodges an electric issue (September 7).

———. 1991a. Foolish ploy dooms Edison deal (November 30).

———. 1991b. Edison case proves need for new law (March 14).

———. 1991c. Private talks and public trust (November 9).

———. 1993. Edison and the great rate war (January 10).

Chubb, John E. 1983. *Interest Groups and the Bureaucracy: The Politics of Energy.* Stanford, CA: Stanford University Press.

Citizens Utility Board (CUB). 1984–1986. *CUB News.* Chicago, Illinois.

———. 1987. Brief of the Citizens Utility Board in the Matter of Commonwealth Edison Company, People of the State of Illinois, Citizens Utility Board, 87-0043, 87-0044, 87-0057, and 87-0096 (May 28).

Clawson, Marion. 1981. *New Deal Planning: The National Resources Planning Board.* Baltimore: The Johns Hopkins University Press.

Committee on Energy and Commerce. 1984. *Hearing before the Subcommittee on Energy Conservation and Power of the Committee on Energy and Com-*

merce, House of Representatives, 98th Congress, 2d session, February 4, 1984. Washington, DC: U.S. Government Printing Office.

Commonwealth Edison Company. Annual Reports from 1984 through 1991.

———. 1987a. Brief of the Commonwealth Edison Company in the Matter of Commonwealth Edison Company, People of the State of Illinois, Citizens Utility Board, 87-0043, 87-0044, 87-0057, and 87-0096 (June 1).

———. 1987b. Reply Brief of the Commonwealth Edison Company in the Matter of Commonwealth Edison Company, People of the State of Illinois, Citizens Utility Board, 87-0043, 87-0044, 87-0057, and 87-0096 (June 5).

———. 1988. Candle, candle, burning bright, could be Chicago every night . . . Chicago: Commonwealth Edison.

———. 1989. Brief of Commonwealth Edison Company in the Matter of Business and Professional People for the Public Interest, et al., v. Illinois Commerce Commission and Commonwealth Edison Company, 68100, 68246, 68247, 68306, and 68355 (Consolidated) (April 25).

Cook, James. 1985. Nuclear follies. *Forbes* (February 11).

Corn, Joseph J., and Brian Horrigan. 1984. *Yesterday's Tomorrows: Past Visions of the American Future.* New York: Summit Books.

Creese, Walter L. 1990. *TVA's Public Planning: The Vision, The Reality.* Knoxville, TN: University of Tennessee Press.

Cronon, William. 1991. *Nature's Metropolis: Chicago and the Great West.* New York: W. W. Norton & Company.

———. 1992. A place for stories: Nature, history, and narrative. *The Journal of American History* 78: 1347–76.

Dalton, Linda C. 1989. Emerging knowledge about planning practice. *Journal of Planning Education and Research* 9: 29–44.

Davidoff, Paul. 1965. Advocacy and pluralism in planning. *Journal of the American Institute of Planners* 31: 331–38.

Davidson, Jean. 1987a. Edison rate-hike war rages on. *Chicago Tribune* (May 23).

———. 1987b. Opponents say Edison's proposed compromises aren't enough. *Chicago Tribune* (June 5).

———. 1987c. Edison's rate increase doomed, ICC says. *Chicago Tribune* (July 2).

———. 1987d. Edison compromise keeps rate plan alive. *Chicago Tribune* (July 7).

———. 1987e. Edison rate plan given new life. *Chicago Tribune* (July 9).

———. 1987f. Edison pulled its own plug. *Chicago Tribune* (July 26).

———. 1987g. Edison seeks 27% rate hike. *Chicago Tribune* (August 22).

———. 1988a. Edison set to deal on rates. *Chicago Tribune* (February 9).

———. 1988b. Industry group hits Edison plan. *Chicago Tribune* (February 12).

———. 1988c. Threat of cuts shocks Edison into rate talks. *Chicago Tribune* (February 14).

———. 1988d. Edison waste estimates vary by $500 million. *Chicago Tribune* (April 11).

Davis, Robert. 1990a. Kidding aside, Daley studies city utility. *Chicago Tribune* (May 9).

———. 1990b. Daley questions Edison's reliability. *Chicago Tribune* (September 5).

———. 1991. City renews takeover threat in Edison talks. *Chicago Tribune* (June 22).

Davis, Robert, and John Kass. 1991. Council clause may mean lights out on Edison pact. *Chicago Tribune* (November 28).

Dear, Michael J. 1986. Postmodernism and planning. *Environment and Planning D: Society and Space* 4: 367–84.

Devall, Cheryl. 1987a. Detractors try to pull plug on funds for Edison study. *Chicago Tribune* (December 28).

———. 1987b. City OKs $400,000 for Edison study. *Chicago Tribune* (December 30).

Dolan, Maureen. 1991. Edison franchise deal needs reworking. Letter to the editor of the *Chicago Tribune* (November 21).

Dold, R. Bruce. 1987a. City studies buying Edison facilities: Proposal could save consumers $18 billion. *Chicago Tribune* (October 16).

———. 1987b. Savings from buyout called a mirage. *Chicago Tribune* (November 5).

———. 1987c. Edison buyout could cost suburbs: Electric bills might rise 17%, consultant says. *Chicago Tribune* (November 4).

———. 1987d. City bid for Edison seen as "serious." *Chicago Tribune* (October 20).

———. 1987e. Rostenkowski flexible on city running utility. *Chicago Tribune* (October 27).

———. 1987f. Two senators back city on Edison. *Chicago Tribune* (December 9).

———. 1987g. Stadiums, Edison buyout shaken: Congress strips away tax breaks. *Chicago Tribune* (December 19).

———. 1988a. Thompson: revive Edison rate-hike plan. *Chicago Tribune* (January 29).

———. 1988b. Edison questions city electric plan. *Chicago Tribune* (January 28).

———. 1988c. City won't buy Edison plants. *Chicago Tribune* (October 19).

———. 1988d. Edison's poll says most oppose buyout. *Chicago Tribune* (October 25).

———. 1989a. Mayor seeks to counterattack Edison in public relations battle. *Chicago Tribune* (January 30).

———. 1989b. City nears decision on Edison franchise. *Chicago Tribune* (August 7).

———. 1989c. Daley wants to end city's Edison pact. *Chicago Tribune* (November 22).

———. 1989d. Daley has 'open mind' about proposal for Edison buyout. *Chicago Tribune* (December 6).

———. 1990. City stalls on Edison franchise. *Chicago Tribune* (May 11).

Dold, R. Bruce, and Michael Arndt. 1988a. Pulling plug on Edison a long shot. *Chicago Tribune* (February 7).

———. 1988b. L.A. offers warning on Edison takeover. *Chicago Tribune* (February 8).

———. 1988c. Edison ignores energy innovations, city charges. *Chicago Tribune* (February 10).

DuBoff, R. B. 1979. *Electric Power in American Manufacturing 1889–1958.* New York: ARNO Press.

Eissman, Mark. 1986a. Consumer advocates don't buy Edison plan. *Chicago Tribune* (December 20).

———. 1986b. ICC leader asks talks on Edison's rate plan. *Chicago Tribune* (December 23).

———. 1986c. Edison won't negotiate on its rate plan. *Chicago Tribune* (December 25).

———. 1986d. Edison reverses, will negotiate $660 million rate increase. *Chicago Tribune* (December 27).

———. 1986e. Edison's hot deal leaves foes cold. *Chicago Tribune* (December 28).

Energy, Environmental Protection and Public Utilities Committee of the Chicago City Council. 1990. In the matter of: Public meeting hearing testimony on the expiration of the contract between Commonwealth Edison and the city of Chicago. Transcript of hearings held on January 1, February 8, 15, and 22, March 1, 7, and 28, 1990.

Energy Policy Project of the Ford Foundation. 1974. *A Time to Choose: America's Energy Future.* Cambridge, MA: Ballinger Publishing Company.

Feld, Marsha Marker. 1989. The Yonkers case and its implications for the teaching and practice of planning. *Journal of Planning Education and Research* 8: 169–75.

Fenn, Scott. 1984. *America's Electric Utilities: Under Siege and in Transition.* New York: Praeger.

Fenno, Richard E., Jr. 1978. *Home Style: House Members in Their Districts.* Boston: Little, Brown and Company.

Ferraro, Giovanni, 1994. "De te fabula narratur." Exercises in reading plans. *Planning Theory* 10–11: 205–36.

Fiddyment, Richard. 1986. *Issues in Utility Resource Planning, Volume I. With An Eye toward the Future: A Discussion of Electric Utility Load Forecasting.* ILENR/RE-SP-86/11–1. Springfield, IL: Illinois Department of Energy and Natural Resources.

Fischer, Claude S. 1982. *To Dwell among Friends: Personal Networks in Town and City.* Chicago: University of Chicago Press.

Fischer, Frank. 1990. *Technocracy and the Politics of Expertise.* Newbury Park, CA: SAGE.

Fischer, Frank, and John Forester, eds. 1993. *The Argumentative Turn in Policy Analysis and Planning.* Durham, NC: Duke University Press.

Fish, Stanley 1979. *Is There a Text in This Class?* Cambridge MA: Harvard University Press.

Fisher, Walter R. 1987, 1989. *Human Communication As Narration.* Columbia, SC: University of South Carolina Press.

Fishman, Robert. 1977. *Urban Utopias in the Twentieth Century: Ebenezer Howard, Frank Lloyd Wright, Le Corbusier.* New York: Basic Books.

Foglesong, Richard E. 1986. *Planning the Capitalist City: The Colonial Era to the 1920s.* Princeton, NJ: Princeton University Press.

Ford, Daniel F. 1981, 1982. *Three Mile Island: Thirty Minutes to Meltdown.* New York: Penguin Books.

Forester, John. 1980. Critical theory and planning practice. *Journal of the American Planning Association* 46: 275–86.

———. 1982. Planning in the face of power. *Journal of the American Planning Association* 48: 67–80.

———. 1989. *Planning in the Face of Power.* Berkeley: University of California Press.

———. 1990. Personal communication.

———. 1993. Learning from practice stories: The priority of practical judgment. In *The Argumentative Turn in Policy Analysis and Planning,* edited by Frank Fischer and John Forester. Durham, NC: Duke University Press.

Fox, Kenneth. 1986. *Metropolitan America: Urban Life and Urban Policy in the United States, 1940–1980.* Jackson: University Press of Mississippi.

Francis, Jere R. 1994. Auditing, hermeneutics, and subjectivity. *Accounting, Organizations and Society* 19, no. 3: 235–69.

Fremon, David. 1988. Commonwealth Edison: A consumer's report. Chicago: Chicago 1992 Committee.

Freund, Elizabeth. 1987. *The Return of the Reader.* New York: Methuen.

Friedmann, John. 1969. Notes on societal action. *Journal of the American Institute of Planners* 35: 311–18.

———. 1973. *Retracking America: A Theory of Transactive Planning.* Garden City, NY: Doubleday.

———. 1987. *Planning in the Public Domain.* Princeton, NJ: Princeton University Press.

Frug, Jerry. 1988. Argument as character. *Stanford Law Review* 40: 869–927.

Funigello, P. J. 1973. *Toward a National Power Policy: The New Deal and the Electric Utility Industry, 1933–1941.* Pittsburgh: University of Pittsburgh Press.

Gadamer, Hans-Georg. 1991. *Truth and Method.* 2d. ed. Translated by J. Weinheimer and D. Marshall. New York: The Crossroad Publishing Company.

Galloway, Thomas. D. and Riad G. Mahayni. 1977. Planning theory in retrospect: The process of paradigm change. *Journal of the American Institute of Planners* 43: 62–71.

Galvan, Manuel. 1987. Council panel supports option of city in any Edison takeover. *Chicago Tribune* (November 3).

Galvan, Manuel, and Rudolph Unger. 1987. Mayor says he'll sue to block quick ruling on Edison rate hike. *Chicago Tribune* (March 20).

Gans, Herbert J. 1962, 1982. *The Urban Villagers: Group and Class in the Life of Italian-Americans.* New York: The Free Press.

Garreau, Joel. 1991. *Edge City: Life on the New Frontier.* New York: Anchor Books.

Garrow, David J. 1986. *Bearing the Cross: Martin Luther King, Jr., and the Southern Christian Leadership Conference.* New York: Vintage Books.

Geertz, Clifford. 1983. *Local Knowledge.* New York: Basic Books.

Gitlin, Todd. 1987. *The Sixties: Years of Hope, Days of Rage.* New York: Bantam Books.

Goldstein, H. A. 1984. Planning as argumentation. *Environment and Planning B: Planning and Design* 11: 297–312.

Goodwin, Craufurd D., et al. 1981. *Energy Policy in Perspective: Today's Problems, Yesterday's Solutions.* Washington, DC: The Brookings Institution.

Gormley, W. T., Jr. 1983. *The Politics of Public Utility Regulation.* Pittsburgh: University of Pittsburgh Press.

Gottlieb, Robert. 1993. *Forcing the Spring: The Transformation of the American Environmental Movement.* Washington, DC: Island Press.

Grabow, Stephen, and Allan Heskin. 1973. Foundations for a radical concept of planning. *Journal of the American Institute of Planners* 39: 106–114.

Grady, William. 1992. Court orders Edison to give refunds. *Chicago Tribune* (April 17).

Grady, William, and Rob Karwath. 1989. Edison loses another rate hike. *Chicago Tribune* (December 22).

Gronbeck, Bruce. 1983. *The Articulate Person.* 2d. ed. Glenview, IL: Scott, Foresman and Co.

Grossman, Karl. 1986. *Power Crazy: Is LILCO Turning Shoreham into America's Chernobyl?* New York: Grove Press.

Habermas, Jurgen. 1987. *The Philosophical Discourse of Modernity.* Translated by F. Lawrence. Cambridge, MA: MIT Press.

Hall, Peter. 1980. *Great Planning Disasters.* Berkeley: University of California Press.

———. 1988. *Cities of Tomorrow.* Oxford: Basil Blackwell.

Hancock, John. 1988. The new deal and American planning: the 1930s. In *Two Centuries of American Planning,* edited by Daniel Schaffer. Baltimore: The Johns Hopkins University Press.

Hardy, Thomas, and R. Bruce Dold. 1988. Bloom tweaks Daley on Edison rates. *Chicago Tribune* (December 23).

Harrington, Michael. 1962. *The Other America: Poverty in the United States.* Baltimore: Penguin Books.

Harvey, David. 1989. *The Condition of Postmodernity.* Cambridge, MA: Basil Blackwell.

Hawkesworth, M. E. 1988. *Theoretical Issues in Policy Analysis.* Albany: State University of New York Press.

Hayden, Dolores. 1984. *Redesigning the American Dream: The Future of Housing, Work, and Family Life.* New York: W. W. Norton & Co.

Haynes, Viktor, and Marko Bojcun. 1988. *The Chernobyl Disaster: The True Story of a Catastrophe—An Unanswerable Indictment of Nuclear Power.* London: The Hogarth Press.

Hays, Samuel P. 1987. *Beauty, Health, and Permanence: Environmental Politics in the United States, 1955–1985.* Cambridge: Cambridge University Press.

Healey, Patsy. 1992. A planner's day: Knowledge and action in communicative practice. *Journal of the American Planning Association* 58: 9–20.

———. 1993. The communicative work of development plans. *Environment and Planning B: Planning and Design* 20: 83–104.

Hellwig, M. 1989. Consumer advocate urges: 'Look at all options.' *One City* 12 (January/February).

Hemmens, George. 1980. New directions in planning theory. *Journal of the American Planning Association* 46: 259–60.

Hendler, Sue, ed. 1995. *Planning Ethics: A Reader in Planning Theory, Practice and Education.* New Brunswick, NJ: CUPR Press.

Hilgartner, Stephen, Richard C. Bell, and Rory O'Connor. 1982. *Nukespeak: The Selling of Nuclear Technology in America.* New York: Penguin Books.

Hill, Edward W. 1989. Yonkers planners acted ethically: Its citizens and politicians acted illegally. *Journal of Planning Education and Research* 8: 183–87.

Hillier, Jean. 1991. Deconstructing the discourse of planning. Paper presented at the Joint ACSP/AESOP International Congress, Oxford, England (July 8–12).

Hirsch, Arnold R. 1983. *Making the Second Ghetto: Race and Housing in Chicago, 1940–1960.* Cambridge: Cambridge University Press.

Hoch, Charles. 1984. Doing good and being right. *Journal of the American Planning Association* 50: 335–45.

———. 1990. Power, planning, and conflict. *The Journal of Architectural and Planning Research* 7: 271–83.

———. 1992. The paradox of power in planning practice. *Journal of Planning Education and Research* 11: 206–215.

Hogan, John. 1986. *A Spirit Capable: The Story of Commonwealth Edison.* Chicago: The Mobium Press.

Houston, Jack, and Jean Davidson. 1987. Advocate blasts Edison rate talks. *Chicago Tribune* (May 17).

Houts, Peter S., Paul D. Cleary, and Teh-Wei Hu. 1988. *The Three Mile Island Crisis: Psychological, Social, and Economic Impacts on the Surrounding Population.* University Park: The Pennsylvania State University Press.

Howe, Elizabeth. 1990. Normative ethics in planning. *Journal of Planning Literature* 5: 123–50.

Hubbard, Preston J. 1961. *Origins of the TVA: The Muscle Shoals Controversy 1920–1932.* New York: W. W. Norton & Co.

Hudson, Barclay. 1979. Comparison of current planning theories: Counterparts and contradictions. *Journal of the American Planning Association* 45: 387–97.

Hughes, Thomas P. 1983. *Networks of Power: Electrification in Western Society, 1880–1930.* Baltimore: The Johns Hopkins University Press.

Hunter, Albert, ed. 1990. *The Rhetoric of Social Research: Understood and Believed.* New Brunswick, NJ: Rutgers University Press.

Hunter, James Davison. 1991. *Culture Wars: The Struggle to Define America.* New York: Basic Books.

Hyman, Lyman. S. 1988. *America's Electric Utilities: Past, Present and Future.* 3d ed. Arlington, VA: Public Utilities Reports, Inc.

Illinois Commerce Commission (ICC). 1971. *Re Commonwealth Edison Co.,* Docket No. 56034, 90PUR3rd 433 (August 13).

———. 1974. In the Matter of General Rate Increase in Electric Rates for Commonwealth Edison Company (April 10). *Utilities Law Reporter, State Decisions 1974–1979.* Chicago, Illinois: Commerce Clearing House.

———. 1975. Application by Commonwealth Edison Company for a Proposed General Rate Increase for Electrical Service (August 27). *Utilities Law Reporter, State Decisions 1974–1979.* Chicago, Illinois: Commerce Clearing House.

———. 1977. Application for a Proposed General Rate Increase in Electric Rates for Commonwealth Edison Company (October 12). *Utilities Law*

Reporter, State Decisions 1974–1979. Chicago, Illinois: Commerce Clearing House.

———. 1978. In the Matter of Authorizing Commonwealth Edison Company a Rate Increase (December 13). *Utilities Law Reporter, State Decisions 1974–1979.* Chicago, Illinois: Commerce Clearing House.

———. 1980a. *In the matter of an investigation of the plant construction program of the Commonwealth Edison Company,* Docket No. 78-0646 (October 15).

———. 1980b. *Re Commonwealth Edison Co.,* Docket No. 79-0214, 35PUR4th 49 (February 6).

———. 1981. *Re Commonwealth Edison Co.,* Docket No. 80-0546, 43PUR4th 503 (July 1).

———. 1982a. *Re Commonwealth Edison Co.,* Docket No. 82-0026, 50PUR4th 62 (May 6).

———. 1982b. *Re Commonwealth Edison Co.,* Docket No. 82-0026, 50PUR4th 221 (December 1).

———. 1984. *Re Commonwealth Edison Co.,* Docket No. 83-0537, 61PUR4th 1 (July 12).

———. 1985. *Re Commonwealth Edison Co.,* Docket No. 83-0537, 84-0555, 71PUR4th 81 (October 24).

———. 1986a. Illinois Public Utilities Act. Springfield, Illinois: Illinois Commerce Commission.

———. 1986b. *Petition requesting the Commission to initiate an expeditious investigation of energy conservation alternatives,* Docket No. 82-0855 (July 2).

———. 1986c. *Re Commonwealth Edison Co.,* Docket No. 80-0706, 77PUR4th 433 (September 17).

———. 1986d. *Re Commonwealth Edison Co.,* Docket No. 83-0035, 78PUR4th 137 (September 17).

———. 1987. *Re Commonwealth Edison Co.,* Docket No. 87-0043, 84PUR4th 469 (July 16).

———. 1988. *Sixth Interim Order: Commonwealth Edison Company proposed general increase in electric rates, et al.,* Docket Nos. 87-0427, 87-0169, 88-0189, 88-0219 consolidated (December 30).

———. 1991. *Re Commonwealth Edison Co.,* Docket Nos. 87-0427, et al. and 90-0169 (March 8).

———. 1993. *Re Commonwealth Edison Co.,* Docket No. 87-0427, et al., 139PUR4th 243 (January 6).

Illinois Commerce Commission Staff. 1987. Brief of the Illinois Commerce Commission Staff in the Matter of Commonwealth Edison Company, People of the State of Illinois, Citizens Utility Board, 87-0043, 87-0044, 87-0057, and 87-0096 (June 1).

Illinois Supreme Court. 1987. The People ex rel. Neil F. Hartigan, Attorney

General, et al., Appellees, v. The Illinois Commerce Commission et al., Appellants. No. 63747. 117 ILL.2d (June 16).

———. 1989. Business and Professional People for the Public Interest et al., Appellants, v. The Illinois Commerce Commission et al., Appellees. Nos. 68100, 68246, 68247, 68306, 68355 consolidated (December 21).

———. 1991. Business and Professional People for the Public Interest et al., Appellants, v. The Illinois Commerce Commission et al., Appellees. Nos. 71602, 71629, 71669, 71681, 71719. 585 N.E.2d 1032 (Ill 1991) (December 16).

Innes, Judith E. 1995. Planning theory's emerging paradigm: Communicative action and interactive practice. *Journal of Planning Education and Research* 14: 183–89.

Ittellag, Richard L., and James Pavel. 1985. Nuclear plants' anticipated costs and their impact on future electric rates. *Public Utilities Fortnightly* 115 (March 21): 35–40.

Jacobs, Jane. 1961. *The Death and Life of Great American Cities.* New York: Vintage Books.

Jacoby, Russell. 1987. *The Last Intellectuals: American Culture in the Age of Academe.* New York: The Noonday Press.

Jenkins-Smith, Hans C. 1982. Professional roles for policy analysts. *Journal of Policy Analysis and Management* 2: 86–99.

Johnson, David A. 1988. Regional planning for the great American metropolis: New York between the world wars. In *Two Centuries of American Planning,* edited by Daniel Schaffer. Baltimore: Johns Hopkins University Press.

Joint Committee on Public Utility Regulation. 1985. *Report of Recommendations to the 84th Illinois General Assembly.* Springfield, Illinois (April).

Joskow, Paul L., and Richard Schmalensee. 1983. *Markets for Power.* Cambridge: MIT Press.

Kahn, Edward. 1988. *Electric Utility Planning and Regulation.* Washington, DC: American Council for an Energy Efficient Economy.

Kaplan, Joel. 1988. Davis urging Edison to open its books. *Chicago Tribune* (December 4).

———. 1989. Edison: only we can light the city. *Chicago Tribune* (April 14).

Kaplan, Thomas J. 1986. The narrative structure of policy analysis. *Journal of Policy Analysis and Management* 5: 761–78.

Karwath, Rob. 1988a. Edison proposes $795 million "modest price" rate increase. *Chicago Tribune* (December 2).

———. 1988b. State OKs 6% Edison rate hike. *Chicago Tribune* (December 22).

———. 1988c. Rate hike approved by Edison. *Chicago Tribune* (December 24).

———. 1989a. Fates of 2 regulators in limbo. *Chicago Tribune* (June 5).

———. 1989b. Time running out in major Edison refund case. *Chicago Tribune* (July 31).

———. 1989c. Thompson set to shake up ICC. *Chicago Tribune* (October 13).

———. 1990a. Franchise fee ruling makes utilities wary. *Chicago Tribune* (April 5).

———. 1990b. Rate law often a bystander in Edison cases. *Chicago Tribune* (January 14).

———. 1990c. Edison wants 17% hike: Rate boost could add $12 to monthly bill. *Chicago Tribune* (April 13).

———. 1990d. Edison to refund consumers about $70 over 3 months. *Chicago Tribune* (June 15).

———. 1990e. Staff at ICC favors Edison rate increase. *Chicago Tribune* (August 25).

———. 1990f. State hearing examiners recommend 7% rate hike for Edison. *Chicago Tribune* (December 19).

———. 1991a. Wording on ballot has Edison fuming. *Chicago Tribune* (March 26).

———. 1991b. State gives Edison 14% rate increase: Consumer groups plan court challenge. *Chicago Tribune* (March 9).

———. 1991c. State high court gets Edison rate dispute. *Chicago Tribune* (April 26).

———. 1992a. ICC chairman Barnich leaves post under fire. *Chicago Tribune* (February 14).

———. 1992b. ICC's Barnich under fire again for calls. *Chicago Tribune* (March 1).

———. 1992c. Edgar nominates 2 to join Illinois Commerce Commission. *Chicago Tribune* (March 12).

———. 1992d. Edison wants ICC rule altered. *Chicago Tribune* (March 4).

———. 1992e. State audit blisters Edison. *Chicago Tribune* (April 11).

———. 1992f. 6.8 percent rate cut is urged for Edison. *Chicago Tribune* (August 11).

———. 1992g. Edison seeks electricity from privately-run firm. *Chicago Tribune* (October 2).

———. 1993a. Ordered to pay fat refunds, Edison says it'll fight. *Chicago Tribune* (January 7).

———. 1993b. Edison taking rate fight to the U.S. Supreme Court. *Chicago Tribune* (January 14).

———. 1993c. To Edison, refund is lesser of 2 evils. *Chicago Tribune* (September 28).

Karwath, Rob, and Michael Arndt. 1990. High court tells Edison it must pay refunds. *Chicago Tribune* (June 1).

Karwath, Rob, and A. Barnum. 1990. W. Side's 2nd blackout called power outrage. *Chicago Tribune* (August 6).

Karwath, Rob, and William Grady. 1991. Edison hike to be reconsidered. *Chicago Tribune* (December 17).

Kass, John. 1991a. Favorable city report on Edison jolts council. *Chicago Tribune* (June 19).

———. 1991b. Daley denies he is a "wimp" to Edison. *Chicago Tribune* (June 20).

———. 1991c. Daley scrambling on Edison. *Chicago Tribune* (June 26).

———. 1991d. City Hall revives talk of buying out Edison. *Chicago Tribune* (Aug 13).

———. 1991e. Key to city Edison deal locked in remap politics. *Chicago Tribune* (November 20).

———. 1991f. Edison kills $70 million city deal. *Chicago Tribune* (December 6).

———. 1991g. Daley gets unions to prod aldermen on Edison pact. *Chicago Tribune* (December 10).

———. 1991h. Gutierrez's shift lets Daley off hook on Edison. *Chicago Tribune* (December 12).

Kass, John, and Robert Davis. 1991a. Edison shocks city on contract demands. *Chicago Tribune* (June 18).

———. 1991b. Daley pulls plug on fight with Edison. *Chicago Tribune* (October 23).

Katz, James Everett. 1984. *Congress and National Energy Policy*. New Brunswick, NJ: Transaction Books.

Kaufman, Jerry. 1990. Forester in the face of planners: Will they listen to him? *Planning Theory Newsletter* 4 (winter): 27–33.

Keating, Ann Durkin. 1988. *Building Chicago: Suburban Developers and the Creation of a Divided Metropolis*. Columbus: Ohio State University Press.

Kihlstedt, Folke T. 1986. Utopia realized: The world's fairs of the 1930s. In *Imagining Tommorrow: History, Technology, and the American Future,* edited by Joseph J. Corn. Cambridge, MA: MIT Press.

Klamer, Arjo, Donald N. McCloskey, and Robert Solow, eds. 1988. *The Consequences of Economic Rhetoric*. Cambridge: Cambridge University Press.

Kleppner, Paul. 1985. *Chicago Divided: The Making of a Black Mayor*. Dekalb: Northern Illinois University Press.

Klosterman, Richard E. 1992. Planning theory education in the 1980s: Results of a second course survey. *Journal of Planning Education and Research* 11: 130–40.

Komanoff Energy Associates. 1990. *Chicago Franchise Options Study: Summary Final Report*. Prepared for the City of Chicago, Department of Planning (May 7).

Kraft, Dave. 1988. Com Ed ads—an exercise in deception? *NEIS News* 7, no. 1: 1–4.

Krieger, Martin H. 1981. *Advice and Planning.* Philadelphia: Temple University Press.

Landow, George P. 1992. *Hypertext: The Convergence of Contemporary Critical Theory and Technology.* Baltimore: The Johns Hopkins University Press.

Langdon, Philip. 1994. *A Better Place to Live: Reshaping the American Suburb.* Amherst: The University of Massachusetts Press.

Lanham, Richard A. 1991. *A Handlist of Rhetorical Terms.* 2d. ed. Berkeley: University of California Press.

Latour, Bruno. 1987. *Science in Action: How to Follow Scientists and Engineers through Society.* Cambridge, MA: Harvard University Press.

LeBel, Phillip G. 1982. *Energy Economics and Technology.* Baltimore: Johns Hopkins University Press.

Leith, Dick, and George Myerson. 1989. *The Power of Address.* London: Routledge.

Lemann, Nicholas. 1991. *The Promised Land: The Great Black Migration and How It Changed America.* New York: Alfred A. Knopf.

Levine, Adeline. 1982. *Love Canal: Science, Politics, and People.* Toronto: D. C. Heath.

Lindblom, Charles. 1959. The science of "muddling through." *Public Adminstration Review* 19: 79–88.

———. 1980. *The Policy-Making Process.* Englewood Cliffs, NJ: Prentice-Hall.

Lindblom, Charles, and David K. Cohen. 1979. *Usable Knowledge: Social Science and Social Problem Solving.* New Haven: Yale University Press.

Loftness, R. L. 1978. *Energy Handbook.* New York: Van Nostrand Reinhold Co.

Lovins, Amory B. 1976. Energy strategy: The road not taken? *Foreign Affairs* 55. no. 1: 65–96.

———. 1977. *Soft Energy Paths: Toward a Durable Peace.* New York: Harper Colophon Books.

———. 1985. Least-Cost Electrical Services as an Alternative to the Braidwood Project: Prefiled Rebuttal Testimony for Petitioners in Illinois Commerce Commission Docket No. 82-0855 and Prefiled Testimony for BPI in Docket No. 83-0035 (July 3).

Lucy, William. 1988. *Close to Power: Setting Priorities with Elected Officials.* Washington DC: American Planning Association.

Lyall, Sarah. 1991. At Shoreham A-plant, a somber beginning of the end. *New York Times* (October 25).

Lynch, Kevin. 1981. *Good City Form.* Cambridge: MIT Press.

Lyotard, Jean-Francois. 1984. *The Postmodern Condition: A Report on Knowl-*

edge. Translated by Geoff Bennington and Brian Massumi. Minneapolis: University of Minnesota Press.

Machiavelli, N. 1977. *The Prince.* Edited and translated by R. M. Adams. New York: W. W. Norton and Co.

McClendon, Bruce W., and Ray Quay. 1988. *Mastering Change: Winning Strategies for Effective City Planning.* Chicago: APA Planners Press.

McCloskey, Donald N. 1985. *The Rhetoric of Economics.* Madison: University of Wisconsin Press.

———. 1990. Storytelling in economics. In *Economics and Hermeneutics,* edited by Don Lavoie. London: Routledge.

———. 1994. *Knowledge and Persuasion in Economics.* Cambridge: Cambridge University Press.

McDonald, F. 1957. *Let There Be Light: The Electric Utility Industry in Wisconsin 1881–1955.* Madison, WI: The American History Research Center.

McGee, Michael C., and John R. Lyne. 1987. What are nice folks like you doing in a place like this? Some entailments of treating knowledge claims rhetorically. In *The Rhetoric of the Human Sciences: Language and Argument in Scholarship and Public Affairs,* edited by John S. Nelson, Allan Megill, and Donald N. McCloskey. Madison: University of Wisconsin Press.

McGee, Michael C., and John S. Nelson. 1985. Narrative reason in public argument. *Journal of Communication* 35: 139–155.

MacIntyre, Alasdair. 1981. *After Virtue.* Notre Dame, IN: University of Notre Dame Press.

———. 1988. *Whose Justice? Which Rationality?* Notre Dame, IN: University of Notre Dame Press.

Maclean, John N. 1991. Franchise talks, rate case to set Edison's course. *Chicago Tribune* (June 16).

———. 1992a. Edison directors to generate a crisis plan. *Chicago Tribune* (July 19).

———. 1992b. Edison slashes jobs, lashes regulators. *Chicago Tribune* (July 23).

———. 1992c. Edison dividend, plants on cost-cutting list. *Chicago Tribune* (July 24).

———. 1992d. Stung Edison to widen energy menu. *Chicago Tribune* (December 10).

Maclean, John N., and Rob Karwath. 1992. Skinner named president of Edison: Job covers regulatory, legal areas. *Chicago Tribune* (December 17).

———. 1993. Edison challenges refunds. *Chicago Tribune* (January 8).

McRoberts, Flynn. 1989. Edison faces $200 million bill. *Chicago Tribune* (August 10).

Majone, Giandomenico. 1989. *Evidence, Argument and Persuasion in the Policy Process.* New Haven: Yale University Press.

Mandelbaum, Seymour. 1988. Open Moral Communities. *Society* 26: 20–27.

———. 1990. Reading Plans. *Journal of the American Planning Association* 56: 350–356.

———. 1991. Telling stories. *Journal of Planning Education and Research* 10: 209–214.

Mandelbaum, Seymour, Luigi Mazza, and Robert W. Burchell. Forthcoming. *Explorations in Planning Theory.* New Brunswick, NJ: CUPR Press.

Markowitz, Paul, and Nancy Hirsh, eds. 1987. *Least-cost Electrical Strategies: An Information Packet.* Washington, DC: Energy Conservation Coalition.

Marris, Peter. 1990. Witnesses, engineers or storytellers: The influence of social research on social policy. In *Sociology in America,* edited by H. Gans. Newbury Park, CA: SAGE.

Marris, Peter, and Martin Rein. 1967, 1982. *Dilemmas of Social Reform: Poverty and Community Action in the United States.* Chicago: University of Chicago Press.

Matthews, Christopher. 1988. *Hardball: How Politics Is Played, Told by One Who Knows the Game.* New York: Harper & Row.

Matthiessen, Peter. 1991. The "Madman" of Chernobyl. *New York Times* (October 14).

Mayer, Harold M., and Richard C. Wade. 1969. *Chicago: Growth of a Metropolis.* Chicago: University of Chicago Press.

Mayor's Task Force on Energy. 1989a. *Recommendations Concerning Electric Energy Policies for Chicago for the 1990s and Beyond.* Chicago: Department of Planning (November).

———. 1989b. Video tapes of public meetings held on April 13, May 2, June 28, and August 16. Chicago: Department of Planning.

Mazziotti, Donald F. 1974. The underlying assumptions of advocacy planning: pluralism and reform. *Journal of the American Institute of Planners* 40: 38–47.

Meadows, Donella H. et al. 1972. *The Limits to Growth: A Report for the Club of Rome's Project on the Predicament of Mankind.* New York: Universe Books.

Megill, Allan. 1989. Pre-postmodernism. Paper presented to the Faculty Rhetoric Seminar of the University of Iowa's Project on Rhetoric of Inquiry. Iowa City, IA: The University of Iowa.

Meltsner, Arnold J. 1976, 1985. *Policy Analysts in the Bureaucracy.* Berkeley: University of California Press.

Meyerson, Martin, and Edward C. Banfield. 1955. *Politics, Planning, and the Public Interest.* Glencoe, IL: The Free Press.

Milroy, Beth Moore. 1991. Into postmodern weightlessness. *Journal of Planning Education and Research* 10: 181–87.

Moberg, D. 1988. One year without Washington. What did he accomplish? How long will it last? *Chicago Reader* 18 (November 25).

———. 1989. Power play: To get a good deal from Commonwealth Edison, the city must wield its buyout clout. *Chicago Reader* 19 (December 1).

Mollenkopf, John H. 1983. *The Contested City*. Princeton, NJ: Princeton University Press.

Moody, Walter D. 1911. *Wacker's Manual of The Plan of Chicago*. Chicago: The Henneberry Company.

Moody's Investor Service. 1967, 1972, 1977, 1984, 1988, and 1991. *Moody's Public Utility Manual*. New York: Moody's Investor Service.

Morone, Joseph G., and Edward J. Woodhouse. 1989. *The Demise of Nuclear Energy? Lessons for Democratic Control of Technology*. New Haven, CT: Yale University Press.

Morris, David. 1982. *Self-Reliant Cities: Energy and the Transformation of Urban America*. San Francisco: Sierra Club Books.

Murray, W. S. 1925. *Superpower—Its Genesis and Future*. New York: McGraw Hill.

Nelson, John S., and Allan Megill. 1986. Rhetoric of inquiry: Projects and prospects. *Quarterly Journal of Speech* 72: 20–37.

Nelson, John S., Allan Megill, and Donald N. McCloskey, eds. 1987. *The Rhetoric of the Human Sciences: Language and Argument in Scholarship and Public Affairs*. Madison: University of Wisconsin Press.

Neustadt, Richard E., and Ernest R. May. 1986. *Thinking in Time*. New York: The Free Press.

Noyelle, Thierry J., and Thomas J. Stanback, Jr. 1983. *The Economic Transformation of American Cities*. Totowa, NJ: Rowman and Allanheld.

Nussbaum, Martha C. 1986. *The Fragility of Goodness*. Cambridge: Cambridge University Press.

———. 1990. *Love's Knowledge*. Oxford: Oxford University Press.

Nye, David E. 1990. *Electrifying America: Social Meanings of a New Technology*. Cambridge, MA: MIT Press.

O'Connell, Mary. 1990. Little city action on electric franchise as December deadline draws near. *The Neighborhood Works* 12: 1.

O'Connor, James J., et al. 1987. Memorandum of Understanding (February 3).

Office of Public Counsel. 1987. Brief of the Office of Public Counsel in the Matter of Commonwealth Edison Company, People of the State of Illinois, Citizens Utility Board, 87-0043, 87-0044, 87-0057, and 87-0096 (June 1).

Payne, B. 1984. Contexts and epiphanies: Policy analysis and the humanities. *Journal of Policy Analysis and Management* 4: 92–111.

People of the State of Illinois. 1987. Brief of the People of the State of Illinois in the Matter of Commonwealth Edison Company, People of the State of Illinois, Citizens Utility Board, 87-0043, 87-0044, 87-0057, and 87-0096 (June 2).

Perin, Constance. 1967. A noiseless secession from the comprehensive plan. *Journal of the American Institute of Planners* 33: 336–47.

Perloff, Harvey S. 1957. *Education for Planning: City, State, and Regional.* Baltimore: The Johns Hopkins Press.

Perrow, Charles. 1984. *Normal Accidents: Living with High-Risk Technologies.* New York: Basic Books.

Phillips, Charles F., Jr. 1993. *The Regulation of Public Utilities: Theory and Practice.* 3d ed. Arlington, VA: Public Utilities Reports, Inc.

Pinkney, David H. 1958. *Napoleon III and the Rebuilding of Paris.* Princeton, NJ: Princeton University Press.

Platt, Harold L. 1987. The Cost of Energy: Technological Change, Rate Structures, and Public Policy in Chicago, 1880–1990. Paper presented at the annual meeting of the Social Science History Association. New Orleans (October).

———. 1991. *The Electric City: Energy and the Growth of the Chicago Area, 1880–1930.* Chicago: University of Chicago Press.

Pringle, Peter, and James Spigelman. 1981. *The Nuclear Barons.* New York: Avon Books.

Quade, Edward S. 1975. *Analysis for Public Decisions.* New York: Elsevier.

Quinn, Arthur. 1982. *Figures of Speech.* Salt Lake City, UT: Peregrine Smith Books.

Rabinovitz, Francine. 1969. *City Politics and Planning.* New York: Atherton Press.

Reardon, Patrick T., and John W. Fountain. 1990. Thousands face another day without power. *Chicago Tribune* (July 30).

Recktenwald, William. 1987. Edison buyout could cut jobs, poll says. *Chicago Tribune* (December 14).

Riis, Jacob A. 1971. *How the Other Half Lives: Studies among the Tenements of New York.* New York: Dover Publications.

Roberts, Marc J., and Jeremy S. Bluhm. 1981. *The Choices of Power: Utilities Face the Environmental Challenge.* Cambridge: Harvard University Press.

Roe, E. M. 1989. Narrative analysis for the policy analyst: A case study of the 1980–1982 medfly controversy in California. *Journal of Policy Analysis and Management* 8: 251–73.

Romo, Robert G. 1989. *The Electric Utility Franchise Expiration and Renewal Process.* Washington, DC: Energy Task Force of the Urban Consortium for Technology Initiatives.

Rorty, Richard. 1979. *Philosophy and the Mirror of Nature.* Princeton, NJ: Princeton University Press.

———. 1989. *Contingency, Irony, and Solidarity.* Cambridge: Cambridge University Press.

Rose, Mark. 1990. *Interstate: Express Highway Politics, 1939–1989.* Rev. ed. Knoxville, TN: The University of Tennessee Press.

Rosenau, Pauline Marie. 1992. *Post-Modernism and the Social Sciences: Insights, Inroads, and Intrusions.* Princeton, NJ: Princeton University Press.

Rosenheim, Daniel. 1985. Utility law changes "hostile," Edison charges. *Chicago Tribune* (April 20).

Ross, Carolyn. 1991. No sense at Edison. Letter to the editor of the *Chicago Tribune* (December 3).

Rudd, David C. 1990. Edison will pay for 3rd blackout. *Chicago Tribune* (September 24).

Rudolph, Richard, and Scott Ridley. 1986. *Power Struggle: The Hundred Years War Over Electricity.* New York: Harper and Row.

Said, Edward W. 1993. *Culture and Imperialism.* New York: Vintage Books.

Sant, Roger W. 1979. *The Least Cost Energy Strategy: Minimizing Consumer Costs Through Competition.* Arlington, VA: Energy Productivity Center.

Sarup, Madan. 1989. *An Introductory Guide to Post-structuralism and Postmodernism.* Athens: University of Georgia Press.

Schneidman, Dave. 1985a. Edison wins 2-step rate boost. *Chicago Tribune* (October 25).

———. 1985b. Edison's Byron 1 plant closed. Chicago Tribune (November 9).

. 1985c. ICC commissioner to be ousted. *Chicago Tribune* (December 13).

———. 1986a. Edison offers rate-hike options for finishing 3 nuclear plants. *Chicago Tribune* (February 28).

———. 1986b. Mayor: Cancel Braidwood plant. *Chicago Tribune* (April 2).

Schneidman, Dave, and Daniel Egler. 1985. Reconfirmed by Senate, commerce commission chief resigns. *Chicago Tribune* (October 16).

Schon, Donald A. 1983. *The Reflective Practitioner: How Professionals Think in Action.* New York: Basic Books.

Schram, Sanford F. 1993. Postmodern policy analysis: Discourse and identity in welfare policy. *Policy Sciences* 26: 249–70.

Schumacher, E. F. 1973. *Small Is Beautiful: Economics As If People Mattered.* New York: Harper & Row.

Scott, Mel. 1969. *American City Planning Since 1890.* Berkeley: University of California Press.

Seigel, Jessica. 1989. Utility watchdog rips 2 picks for ICC. *Chicago Tribune* (October 23).

Simons, Herbert W., ed. 1989. *Rhetoric in the Human Sciences.* London: SAGE.

———, ed. 1990. *The Rhetorical Turn: Invention and Persuasion in the Conduct of Inquiry.* Chicago: University of Chicago Press.

Solar Energy Research Institute. 1981. *A New Prosperity: Building a Sustainable Energy Future.* Andover, ME: Brick House.

Squires, Gregory D., Larry Bennett, Kathleen McCourt, and Phillip Nyden. 1987. *Chicago: Race, Class, and the Response to Urban Decline.* Philadelphia: Temple University Press.

Stern, Paul C., and Elliot Aronson, eds. 1984. *Energy Use: The Human Dimension.* Committee on Behavioral and Social Aspects of Energy Consumption and Production, Commission on Behavioral and Social Sciences and Education, National Research Council. New York: W. H. Freeman and Company.

Stewart, Charles, Craig Smith, and Robert E. Denton, Jr. 1984. *Persuasion and Social Movements.* Prospect Heights, IL: Waveland Press.

Stokey, Edith, and Richard Zeckhauser. 1978. *A Primer for Policy Analysis.* New York: Norton.

Stone, Susan B. 1988. Concurring/Dissenting Opinion to the Sixth Interim Order entered by the Commission on December 30, 1988. Springfield: Illinois Commerce Commission.

Strong, James, and Stanley Ziemba. 1986. Washington renews threat to pull plug on Edison. *Chicago Tribune* (August 6).

Susskind, Lawrence, and Jeffrey Cruikshank. 1987. *Breaking the Impasse: Consensual Approaches to Resolving Public Disputes.* New York: Basic Books.

Sussman, Carl. 1976. *Planning the Fourth Migration: The Neglected Vision of the RPAA.* Cambridge, MA: MIT Press.

Tanzman, Edward A. 1990. Personal communication.

Tarr, Joel A., and Gabriel Dupuy, eds. 1988. *Technology and the Rise of the Networked City in Europe and America.* Philadelphia: Temple University Press.

Tett, Alison, and Jeanne M. Wolfe. 1991. Discourse analysis and city plans. *Journal of Planning Education and Research* 10: 195–200.

Throgmorton, J. A. 1989. Synthesizing politics, rationality, and advocacy: Energy policy analysis for minority groups. *Policy Studies Review* 8: 358–67.

———. 1990. Passion, reason, and power: Electric power planning in Chicago. *Journal of Architectural and Planning Research* 7: 330–50.

———. 1991. The rhetorics of policy analysis. *Policy Sciences* 24: 153–79.

———. 1992. Planning as persuasive storytelling about the future: Negotiating an electric power rate settlement in Illinois. *Journal of Planning Education and Research* 12: 17–31.

———. 1993. Planning as a rhetorical activity: Survey research as a trope in arguments about electric power planning in Chicago. *Journal of the American Planning Association* 59: 334–46.

Throgmorton, J. A., and Martin Bernard III. 1986. Minorities and energy: A review of recent findings and a guide to future research. *Proceedings of the American Council for an Energy-Efficient Economy's 1986 Summer Study on*

Energy Efficiency in Buildings Vol. VII. Washington, DC: American Council for an Energy-Efficient Economy.

Throgmorton, J. A., and Peter S. Fisher. 1993. Institutional change and electric power in the city of Chicago. *Journal of Economic Issues* 27: 117–53.

Tompkins, Jane P., ed. 1980. *Reader-Response Criticism: From Formalism to Post-Structuralism.* Baltimore: The Johns Hopkins University Press.

Toulmin, Stephen. 1990. *Cosmopolis: The Hidden Agenda of Modernity.* New York: The Free Press.

Unger, Rudolph. 1988. State commission proposes Edison hike. *Chicago Tribune* (June 7).

U.S. Department of Energy. 1981. *National Electric Reliability Study: Executive Summary.* DOE/EP-0003. Washington, D.C.: U.S. Government Printing Office.

———. 1990. *Annual Energy Review.* DOE/EIA-0384 (90). Washington, D.C.: U.S. Government Printing Office.

U.S. Office of Technology Assessment. 1984. *Nuclear Power in an Age of Uncertainty.* OTA-E-216. Washington, DC: U.S. Government Printing Office.

Urbanism Committee of the National Resources Committee. 1937. *Our Cities: Their Role in the National Economy.* Washington, DC: U.S. Government Printing Office.

Warren, Karen J. 1987. Feminism and ecology: Making connections. *Environmental Ethics* 9: 3–21.

Weart, Spencer R. 1988. *Nuclear Fear: A History of Images.* Cambridge: Harvard University Press.

Weinstein, Natalie, and Sharman Stein. 1990. Utility bill breaks for outages urged. *Chicago Tribune* (August 10).

Wetlaufer, Gerald B. 1990. Rhetoric and its denial in legal discourse. *Virginia Law Review* 76: 1545–97.

White, James Boyd. 1984. *When Words Lose Their Meaning: Constitutions, and Reconstitutions of Language, Character, and Community.* Chicago: University of Chicago Press.

———. 1985. *Heracles' Bow: Essays on the Rhetorics and Poetics of the Law.* Madison: University of Wisconsin Press.

Wildavsky, Aaron. 1979, 1987. *Speaking Truth to Power: The Art and Craft of Policy Analysis.* New Brunswick, NJ: Transaction Books.

Williams, Charles. 1990. Interview with author.

Wilson, Elizabeth. 1991. *The Sphinx in the City: Urban Life, the Control of Disorder, and Women.* Berkeley: University of California Press.

Wingert, Pat, and Hanke Gratteau. 1985. Edison rate increase ignites governor's race. *Chicago Tribune* (October 28).

Wrigley, Robert L., Jr. 1983. The plan of Chicago. In *Introduction to Planning*

History in the United States, edited by D. A. Krueckeberg. New Brunswick, NJ: Center for Urban Policy Research.

Yannow, Dvora, 1993. The communication of policy meanings: Implementation as interpretation and text. *Policy Sciences* 26: 41–61.

Ylisela, James, Jr., and Lisa Capitanini. 1991. Edison's reliability is still in doubt: Secret pact with city hides answers. *The Chicago Reporter* 20, no. 6 (June): 1.

Ziemba, Stanley. 1986. Edison rates may cut jobs: City study assails proposed increases. *Chicago Tribune* (August 5).

Ziman, John. 1984. *An Introduction to Science Studies: The Philosophical and Social Aspects of Science and Technology.* Cambridge: Cambridge University Press.

Illustration Credits

1	Photo by Derek Maurer
1.1	Courtesy Chicago Historical Society (ICHi-02554)
1.2	© 1996 Artists Rights Society (ARS), New York/SPADEM, Paris
1.3	Gift of Helen Farr Sloan. © 1995 Board of Trustees, National Gallery of Art, Washington
3.1	Courtesy Moody's Investors Service
4.3	Courtesy Commonwealth Edison
4.4	Courtesy Commonwealth Edison
4.5	Courtesy Illinois Commerce Commission
5.1	Chicago Tribune photo by Michael Fryer
5.2	Chicago Tribune photo
5.3	Chicago Tribune photo by Chris Walker
5.4	Chicago Tribune photo by Gerald West
5.5	Chicago Tribune photo by Steve Hanks
6.1	Chicago Tribune photo by Don Casper
6.2	Chicago Tribune photo by Chris Walker
6.5	Courtesy Illinois Commerce Commission
7.1	Chicago Tribune photo by Val Mazzenga
7.2	Chicago Tribune photo by Chris Walker
7.4	Paul Hosefros/NYT Pictures
7.5	Courtesy Commonwealth Edison
7.6	Chicago Tribune photo by Michael Fryer
9.1	Steve Kagan/NYT Pictures
9.2	Chicago Tribune photo by Jose More
9.3	Chicago Tribune photo by Chris Walker
10.1	Courtesy Illinois Commerce Commission
10.2	Courtesy Commonwealth Edison
10.5	Chicago Tribune photo by David Klobucar
11.2	Chicago Tribune photo by Walter Kale

Index